Career in Optics

A Memoir that Spans 36 Years (1950-1986) of Optics History at Eastman Kodak

PIERCE BURTON DAY

"Mr. Day, more than anyone else, can be considered the father of the K optical system..."

— R.E. Kleim, R.H. Hollenbeck, R.W. Tietge, L.T. Askren

The Day Group

Career in Optics
A Memoir that Spans 36 Years (1950-1986) of Optics History at Eastman Kodak

Published by: **The Day Group, Inc.**
Two Union Square
601 Union Street, Suite 4200
Seattle, WA 98101-4036

Printed in the United States of America

"Career in Optics" begins at the University of Rochester, Institute of Optics, followed by a year at NACA, the predecessor to NASA, in Cleveland, Ohio (1948-1949) and 36 years (1950-1986) at Eastman Kodak. These years spanned a "Golden Era" — when optics of every shape and size were made at Hawk-Eye and when Kodak pioneered the development of the lenses, the film and the cameras for the U.S. secret satellite reconnaissance programs. It also documents the development of optics for the Ektaprint copier-duplicators.

Cover Image: An Atlas missile lift-off with an Agena-B second stage
Cover Design: William Sabia
Book Design & Editing: James P. Day

Library of Congress Cataloging-in-Publication Data
LCCN: 2 0 1 4 9 4 0 0 2 6

Day, Pierce B., 1924 -
Career in Optics / Pierce. B. Day
p. cm.
Includes bibliographic references and footnotes
ISBN 978-0-692-21077-2 (Soft Cover)

Dedication

To the memory of those venerable Institute of Optics professors who engaged the minds and imagination of their students — opening the windows to unimagined worlds of optics:

Brian O'Brien, M. Parker Givens, Robert E. Hopkins

and

Rudolf Kingslake

who put me on track and launched my Kodak career in optics.

Table of Contents

Acknowledgements

I wish to acknowledge a deep sense of indebtedness to the University of Rochester's Institute of Optics for the unique and rewarding career in optics that resulted in large measure from choosing optics as an undergraduate. I feel especially privileged to have studied under the Masters. Pioneers who established the Institute as a world-class center for optics: Brian O'Brien, Robert Hopkins, Parker Givens and Rudolf Kingslake. They inspired a life long fascination with all things optical starting with the fundamental concepts in those arcane disciplines: physiological, physical and geometrical optics. They opened the window to the world of optical science that would lead in its application to placing America's "Secret Eyes in Space." I was tethered to that world throughout my career. It could not have happened without Kingslake's totally unexpected offer of a position in the lens design department at Kodak.

Without question Kingskake's oversight of the complete range of optics development, including programs affecting national security, was of incalculable value to Kodak and I benefitted by reputation as his protégé.

Two others were instrumental in mentoring my professional development: Gordon Milne, who saw me through my masters project, and Fred Wolff, a fraternity brother who preceded me at Kodak and who, quite serendipitously, became my boss, for his editorial assistance during my early struggles with the preparation of coherent engineering reports.

Preface

As one is said to be advanced in age at 80, and advancing at an alarming rate, one's thoughts inevitably turn to down-sizing, deciding what to dispose of and what to preserve of life's accumulated treasures. For some reason when I retired in October of 1986, I took with me a small library of books and pamphlets, plus a box of files representing the remnants of a career in optics. Perhaps I thought I might be recalled by Kodak or would take on a consulting job. It never happened and my professional life as an optical engineer receded into long term memory, that is until recently when I began to delve into the files again. I realized that my career in optics spanned the best years of Kodak as an optics and camera manufacturing company and if I wrote my story it would coincide with the rise and fall of Kodak as a world-class optics manufacturer.

Kodak was also world-class as a company to work for, as employees of my generation would attest. I felt a sense of obligation to write this story not only because of my career at Kodak but also because of those with whom I worked and because of the Institute of Optics at the University of Rochester, the institution that made it all possible. So that's where the story begins — the years returning to the U of R after the war.

I graduated in 1948 with a B.S. in optics and accepted a job in Cleveland, Ohio (1948-1949) at the national research facility for aeronautics, NACA, which later became NASA. My career in optics continued at Kodak (1950-1986). Roughly half of those 36 years at Kodak were in the Research and Engineering (R&E) Division where I was involved in the U.S. intelligence community's satellite reconnaissance programs. They are documented in the public record only to the extent that information has been declassified or has leaked through the cracks in the nation's security firewall.

A considerable volume of conjectural, filtered information exists on the internet and in several books the most notable of which is the definitive work of Jeffrey Richelson, "America's Secret Eyes in Space." At first I read the history in the public record as a curious bystander but then I was drawn into it as a participant, curious as to how my part in it turned out. To be part of a great enterprise and denied access to whether it was a success or failure left an emptiness. For years it was a disappointment not knowing, always this sadness in looking back at the most memorable years of my career at Kodak, looking for resolution.

It was all there in Richelson's book the deeper I got into it, past the politics there were the answers I was looking for — the numbers that, for me at least, told the story of success and the words that I wanted the world to hear:

> "History may seem like a trivial concern to some, but it's the basis for understanding the past and acting in the future. The history of the KEYHOLE programs is a history of one of the most significant military technological developments of this century and perhaps in all history."

Hyperbole or not, the history in the public record, in my view, was flawed, incomplete and without some accounting of the breakthroughs in technology that led to success and contributed to the more openly reported successes of the Hubble and Apollo programs.
So there was the motivation to fill in the gaps and to the best of my limited ability, set the record straight. Richelson makes the point, especially valid in my case, that secrecy has severely limited access to individuals involved in classified programs.

> "And those individuals are not immortal. Many of those involved in the early programs have passed away. By the year 2013, several more pioneers will be gone."

To those students at the Institute of Optics looking for a career in optics, I can only say that what you have learned is a priceless asset not only in finding work but also for the country in dire need of scientists to push the boundaries. I wish you all the success.

PBD
Pittsford, N.Y.
2013

Pierce
DAY
171 Raleigh Street
OPTICS

1

The Institute of Optics
University of Rochester

With the defeat of Germany in Europe and Japan in the Pacific, the future for America held unlimited promise. The mood everywhere of enterprising optimism energized and unified the entire nation. The great military industrial complex that supplied the troops, delivered the tanks, the planes and the bombs could now be dismantled (or at least so we thought) and converted to reconstructing a peacetime economy.

Back then there were men of conscience and extraordinary vision leading the nation. They saw that to secure the peace, war-torn Europe had to be restored. With bipartisan support the Marshall Plan for the reconstruction of Europe passed through Congress and was signed into law by President Truman.

Equally important to the nation, as well as to those of us who survived the war, was the GI Bill which helped millions of servicemen to a free education. Colleges throughout the country were inundated.

Living Off-Campus

Many of the GIs returning to the University of Rochester (UofR), were married, bringing their war brides with them. Some were lucky enough to have signed up for one of the apartments in the new housing units hastily constructed when the war ended. Others were left to fend for themselves hoping to find housing in the surrounding neighborhoods. Some had already started families, but others considered discretion the better part of valor, putting off having children to after graduation. So the disciplines of military life carried over to the serious business of acquiring the professional skills and knowledge required to support a family.

We were no different, my wife and I, for those first two years of marriage. We lived in the tiny third floor attic apartment at 171 Raleigh Street, a block over from Crittenden Boulevard. The novelty of living together that summer was only the beginning of a new life so far removed from the dehumanizing regimentation of army life. We felt released from the memories, the frustration of life put on hold and now looked to a bright future. With her New York State nurse's license, my new bride could work at Strong Memorial Hospital, only a ten minute walk from the apartment. For me it was a ten minute bike ride to campus.

I had to have a job that summer to augment what Sally could earn at 6 dollars a day at the hospital, since the $120 a month that we would get from the G.I. Bill didn't start until September. It was with the RT French Company unloading cement blocks from a flat car to a wheelbarrow and running them over to the construction site. It was a good thing I had a strong back then but it was a hot summer, over 90 degrees, and I would take a short break between loads. It was just at one of those times that the big boss saw me just standing there and fired me on the spot. There was no such thing as a minimum wage so the loss of pay was not nearly as painful or as

Kneeling, Left to Right: *Coach Campbell, K. Bastress, P. Woods, J. McHugh, R. Koch, P. Day, A. Gilman, R. Kanka, J. Ballarian, R. Currie.* ***Standing, Left to Right:*** *W. Britton, T. Mapp, A. Van der Laande, J. Pasceral, W. Hamlin, C. Wadhams, P. Svanoe, T. Searle, D. Kock, R. Blakeslee, Mgr. F. Hammil.* ***Absent:*** *R. Gray, K Holand. M. Battle* ***Record:*** *Rochester 4 - Ithaca State 3; Rochester 3 - Bufffalo State 1; Rochester 2 - Colgate 0; Rochester 6 - Allegheny 1; Rochester 3 - Hamilton 1; Rochester 0 - Cortland 1; Rochester 2 - Syracuse 1*

Photograph 1: **University of Rochester Soccer Team**

Coach "Doc Campbell " looked over his opening soccer turnout and opined that the 1947 club might be a pretty fair team. "The father of Rochester soccer" was right, the 1947 squad went on to pile up the finest of any Uof R soccer teams in history with six wins and one loss. The Rivermen got off on the right foot, trimming Ithaca State, 4-3, as Kirk Howland and Pierce Day got two goals apiece. The Cambellmen made it four in a row when they pasted hapless Alleghany, 6-1, as left inside Day accounted for three goals.

Standing, Left to Right: *A. Gourley, C. Cochrane, J. Krieger, B. Lovell, L. Block, D. Fisher, W. Ryan, P. Day, Coach Campbell*

Photograph 2: **Tennis Team**

Under the leadership of Bob Lovel at the number one position, Wally Ryan, Rick Raible, Pierce Day, Len Bloch, John Dieterle and Ted Weeks took their respective positions. Singles matches were played by the first six men and the doubles by Lovell and Day, followed by Raible and Ryan.

embarrassing as facing Sally at the end of the day.

Shortly after that, I had a job as a lab assistant in the Optics Department. It was my first exposure to the new-at-the-time science of thin-film coatings on glass. I was impressed by the young researcher who had her own lab fully equipped with a vacuum chamber and all sorts of optical test equipment. As I learned later, Mary Banning was the daughter of the famous author, Margaret Culken Banning, who had a Ph.D. from Johns Hopkins. In 1947 she published a ground-breaking paper on how to deposit and control the thickness of multilayer filters. I mention this because years later I would become intimately involved with the specification and procurement of these filters.

The excitement of being back on campus with a full schedule of classes was only the half of it. I couldn't wait to get back to playing soccer, playing the horn in band and stopping by the fraternity. It was all like it was before the war except at the end of the day I would hop on my bike and ride off to my little attic paradise.

Academically Challenged

Soccer was a great diversion from the academically challenging schedule at the Institute of Optics. It was also a great body builder and self-confidence builder, especially after we started winning games. I was scoring goals in most of the games — two at Ithaca and three at Allegheny. It had been over two years away from the rigors of academia, and my schedule was loaded with subjects entirely new to me, like Thermodynamics and Chemistry, which placed a tremendous burden on the memory — especially Chemistry. I did poorly that first semester and dropped it, since I could see no connection with Optics. One of my most fascinating courses was Physiological Optics, taught by the renowned professor and director of the Institute of Optics, Brian O'Brian. I couldn't have had a more stimulating introduction to the science of optics.

It was all about the anatomy of the eye, how it responds to light and how we perceive images of the outside world. I was quite amazed by the eye's ability to keep the outside world, near and far objects, in focus without our even thinking about it. The eye, in effect, adapts to the quality or sharpness of the image on the retina in a process called accommodation. If the eye does this perfectly, the eyeball is said to be emmetropic. But more often than not the eyes become myopic or hyperopic — defects that may develop over time because the lens focal length and the length of the eyeball get out of whack. Then we have the most interesting condition of all that has nothing to do with being Presbyterian. It's a condition called Presbyopia which is a loss of accommodation as you get older. So you might say that us older Presbyterians are truly far-sighted! Rather than go on with this tutorial, I should mention two other troublesome defects of the eye, as you get older, glaucoma and cataracts, which we learn about from experience.

A Degree in Optics

As an undergraduate I had absolutely no idea of the job opportunity there might be after graduation or whether the courses I took would be relevant. I was only vaguely aware of the important contributions of Brian O'Brian and others to the war effort. That was all behind me. But the Institute of Optics was acquiring a reputation as a center for research and applied optics. It would soon become the heart of the burgeoning optics industry in Rochester as well as the country. In fact two of my professors were lured into industry, Brian O'Brian to American Optical and Rudolf Kingslake to Kodak as Head of the Lens Design Department, but would maintain their ties to the university.

So it was my incredible good fortune to be there at the right time. Perhaps the most important course from the standpoint of shaping my career in optics was Kingslake's lens design and geometrical optics.

In the fall of '47, I had the choice of concentrating my studies in lens design and geometrical optics or physical optics- a more theoretical, "pure" science. I chose the latter more out of curiosity than any insight. It turned out to be a fortuitous decision in several ways. The Institute had just made the appointment of M. Parker Givens, a Ph.D. from Cornell, to develop the course in physical optics and I had the privilege of being one of his first students. I believe most of the course was devoted to theories about the nature of light and as with all scientific theories, there were the mathematical formulations to describe certain phenomena. So I was introduced to the distinguished scientists in the field who had formulated equations for the coherence and interference properties of light (A.A. Michelson, Max Von Laue, E. Schrodinger, and F. Zernike — all Nobel Prize winners). I must have been proficient in the math but none of it stuck very long probably because I never had occasion to use it.

However, a lab experiment would vividly demonstrate both the coherence and the interference of light and my fascination would extend into a career applying and promoting interference techniques. With a simple arrangement of mirrors and a beam-splitter, called a Michelson interferometer, two coherent beams of monochromatic light are separated and brought back together to produce light and dark interference fringes. If one of the beams is bounced off a perfect mirror, for example, the fringes are perfectly straight and any defect in the mirror could be measured in wavelengths of light. At the time interference testing of optical systems or mirrors was not widely used in industry.

My impression of Parker Givens, as a laid-back, unassuming man of uncommon intelligence, still remains after fifty or more years. As a friend he instilled the desire and love of learning in his students and for me he opened up not only the world of optics but the universe. The way astronomers catalogue the size or magnitude of the stars and measure the distance to the stars in light-years fascinated me. Its hard to grasp the size of the universe until you know that light travels at 186,400 miles per second and multiply that by the number of seconds per year to get the number of miles per light-year. Now that you have a big enough yardstick, you can begin to comprehend distances to the stars that range from ten to a hundred million light-years away. Talk about stretching the imagination.

The course in lens design and geometrical optics was taught by Rudolf Kingslake who, by 1947, was already nationally known as head of the Lens Design Department at Eastman Kodak. But at the time I was clueless. I never thought of working at Kodak or that Kingslake would play such a pivotal roll in my career. I didn't know what to think of this tall rather imposing English gentleman — perhaps a combination of respect and fear that I could never measure up.

We didn't know the Kingslakes socially until we came back to graduate school but I knew that his wife, Hilda, was the daughter of a Professor Conrady at Imperial College, England, who wrote the book on lens design. It was not a text that you could pick up and read easily as it was far too advanced. I was sufficiently impressed by the complexity of the math that I realized I probably would not be a lens designer. However, I did learn about the basic concepts of lens-design which involves the correcting or rather minimizing of the various aberrations over some specified field of view.

The formulas for ray tracing were quite straightforward. In the lab we learned to use a Marchant calculater for tracing rays through a lens (actually a system of lens elements) surface by surface — a tedious, boring procedure. But the end result was a spot diagram (when plotted out) of where the rays intersect the image plane which is kind of fascinating. This was a purely artificial look at the image of a point source. Then, with a lens set up on a lens bench, we looked at real images in Technicolor. I forgot to mention that in addition to the monochromatic aberrations of spherical, astigmatism and coma, there are all the chromatic aberrations. Well, enough of that!

Bob Hopkins was the instructor. With a B.S. from M.I.T. and having just received his Ph.D. from the Institute of Optics, he was well qualified as a mentor in learning about various lens types, binoculars and telescopes. I'm sure it was there that I discovered the beauty of the Galilean telescope. This is where the length of the telescope could essentially be cut in half. Some years later I would actually be in a periscope design competition with Bob Hopkins. Galilean telescopes would become the lenses launched into space.

With only fourteen in the class of '48 there was a close relationship between faculty and students and for quite a few years I looked forward to meeting optics graduates at reunions. Vance Carpenter, Donald Cary and George Hart were also fraternity brothers. Don Koch and Bob Rosborough were lifelong friends together with their wives. Then there were a number of graduates who ended up at Kodak or Tropel who made their mark in the optics industry.

2

The First Job
National Advisory Committee for Aeronautics

In May of 1948 the college years were rapidly coming to a close. My wife and I would be leaving behind an unforgettable part of our lives. The parties at the fraternity and Spiegles, the tennis matches, the cramming for exams, and of course, the cozy attic apartment at 171 Raleigh Street.

It had to end and we faced the frightening prospect of having to look for a job. With the GI Bill and my stipend from the Presbyterian Board of Foreign Missions ending at graduation, I would soon be without any visible means of support.

At some point I had accepted an offer from the National Advisory Committee for Aeronautics, a government lab in Cleveland, Ohio. It was the predecessor to the National Aeronautics and Space Administration (NASA). Whether I was just too busy getting ready to leave or had some anti-establishment, independent attitude, I decided to skip the commencement exercises. At any rate, it was pretty hectic those last days and it was sad saying good-byes to all our friends. We had to buy train tickets and reserve a hotel room in Cleveland. The one last thing I had to do was pick up a varsity sweater — a white cardigan with the letter "R". Only three-letter-men were entitled to the white sweater. I already had two of the blue varsity sweaters so this one was something special!

Olmstead Falls

Answering an ad for a 2-room flat with a private entrance, we drove to the lovely little college town of Olmstead Falls south west of Cleveland and only ten minutes from the Lab.

The house was a large two-story Swiss Chalet set apart from neighbors on a kind of secluded lot. In the back were two rentals, a larger apartment on the ground level and the two room flat above it which you got to by climbing these wooden steps (actually a fire escape as mandated by law).

The landlady, a Mrs. Geist, was a large friendly woman who took us through her home to show us the rooms. We were delighted and said we had no problem sharing the bathroom. She was quite pleased, in fact, when I said I would like to paper the bedroom. The furnishings were the bare minimum. In the bedroom there was simply a bed, chest of drawers, and a closet. The living room had a couch, a stuffed chair and drop-leaf table with two chairs placed by a big window that let in the morning sun. Lined up against the interior wall were the refrigerator, a good-sized sink and a two-burner table-top stove with a toaster oven. Later Sally's mother sent us an electric oven that we had for years.

So this was our second cozy apartment high up and isolated from the world. It was just the two of us, but not for long. In fact, we had company, in-utero. Sally was pregnant.
On May 12th a Friday, her water broke, as they say, and we rushed her to Lakewood Hospital.

Thirty-six hours later on May 14th Sally delivered, or was delivered of, a 9 pound 5 ounce baby boy. We named him James Pierce.

National Advisory Committee for Aeronautics (NACA) — The Lewis Flight Propulsion Laboratory

The Lewis Lab in Cleveland was one of three government research facilities originally established to support the aircraft industry. The Cleveland lab was already a large complex of buildings including a jet engine test facility and the country's largest wind tunnel. It was absolutely huge if you can imagine the scale of a tunnel at least eight feet in diameter and two city blocks long. Gigantic fans at one end could move air through the tunnel approaching the speed of sound, simulating the airflow over the wings of a jet. Residents on the west side knew when the fans were turned on.

So why in the world would they hire someone just fresh out of the Institute of Optics. I don't remember the total employment, probably in the thousands, but I was to assist the one employee with any knowledge of optics, a physicist by the name of Donald Buchele, a likeable fellow with a sharp mind. I enjoyed working for him. However, the working environment left something to be desired. Instead of the plush offices in the main building, Buckley preferred space directly over the jet engine test chambers. There were two of them so that when one side was down, a jet engine was running at full throttle on the other side. The noise level was deafening but you got used to it. It certainly would have been in violation of EPA work-place noise level standards, if the EPA had been in existence.

When he took me over to the wind tunnel, I was amazed at the sheer size of the test chamber and the Schlieren optical system that our optics group of two was responsible for. Buchele had it designed and installed. Some optics shop in Toledo had ground and polished the two massive six foot diameter mirrors mounted on opposite sides of the test chamber. They were off-axis aspheres to eliminate spherical aberration and provide a perfectly parallel light-beam, passing through glass windows in the sides of the tunnel. Light from a very intense source at the focus of the first mirror produced a "collimated" beam, which was then picked up and focused by the second mirror on a small aperture. With this arrangement a picture, like a shadow-graph, could be made of the air-flow over a wing cross-section. You could see the boundary layer and turbulence at the trailing edge of the wing but air density measurements were only approximate.

Commercial Jets were still on the drafting boards at Boeing and engineers needed accurate, quantitative data to determine the shape of the wing for maximum "lift". I'm not sure when the wheels started turning in my head but I began thinking along the lines of converting from a Schlieren system to what is technically referred to as a Mach-Zender interferometric system. This was right down my alley but a tremendous challenge. The Mach-Zender approach was totally impractical. The situation called for an invention. They say that "Necessity is the Mother of Invention" and I suppose I nursed this idea for the next several months.

Photograph 3: **Aerial View of Glenn Reseach Center at Lewis Field (Cleveland, Ohio)**
An aerial view of the NASA John H. Glenn Research Center, formerly the Lewis Flight Propulsion Laboratory — one of three National Advisory Committee for Aeronautics (NACA) facilities. (NASA image from Wikipedia website.)

Photograph 4: **Supersonic Wind Tunnel at NACA's Lewis Lab**
The supersonic wind tunnel at the Lewis Flight Propulsion Lab as it was called in 1944 when the picture was taken. I worked on the second floor of the building on the right directly above the two jet engine test chambers. (Image from Wikipedia website.)

The Dead-end Job

Meanwhile I was becoming disillusioned with my civil service job. We were barely getting by on my meager $240 a month (Annual salary: $3,351) and I could see that advancement would come in small increments over a period of years. I had responsibilities to a family now with no guarantee that it wouldn't increase in size. It was a dead-end as far as my future in optics was concerned. Since I had one year of eligibility left on the G.I. Bill, I decided I had to go back to the UofR for my Masters. Buckley's boss, Skoog, wasn't too happy when I gave my notice.

The "Differential Interferometer"

I had come up with the idea to convert the large wind tunnel Schlieren system over to an interferometer. Since there was no way to pass the reference beam around the tunnel, I thought a Shiering type interferometer might work but I never referred to it as such in my invention report. At least the configuration of mirrors may have been novel.

A patent application was submitted, jointly with Donald Buchele, for a "Differential Interferometer"— a combination of terms unique to the patent literature. Once the paper mill was set in motion, the patent attorneys were happily occupied for about a year. I had forgotten all about it.

An Inconvenient Breakdown

We were leaving Cleveland with few regrets and high expectations for the future. We arrived, just the two of us, traveling by bus and now we had wheels traveling with all our worldly possessions packed in the trunk and a three-month-old Jimmy tucked snuggly in the car bed behind our seats.

Our first stop was Fredonia to introduce Jimmy to his Uncle Jim and Aunt Gerry (Briggs). Never having had children of their own, the Briggs welcomed Jimmy as their own first grandson. It was late August and we slept on the upstairs, screened porch with a summer breeze and Jimmy just inside in his crib. And so it was that the Briggs's beautiful home on the hill in Fredonia became our second home and they watched and reveled in our growing family as we returned year after year.

After waving goodbye to the Briggs, we were cruising along Route 20 just outside of Hamburg, NY, when our here-to-fore smooth running Studebaker started having convulsions and with a sudden jolt and a clickity-clack it threw a rod right up through the block — a real catastrophe! What would we do? I had to be in Rochester to register and we couldn't possibly get there by bus. I was stymied. As is often the case in emergencies, women have a clearer head. Sally simply called Uncle Jim who came in the big Buick to pick them up. Somehow I got the car towed to a garage, arranged to pick it up in a few days and proceeded on by bus.

In Rochester I was able to take care of business on campus and then start looking for a place that would accommodate our small family and our rather limited budget. We would still have $120 a month from the G.I. Bill.

NATIONAL ADVISORY COMMITTEE
FOR AERONAUTICS
1512 H STREET, NORTHWEST
WASHINGTON 25, D. C.

TELEPHONE: LIBERTY 5-6700

October 12, 1954

REGISTERED MAIL

Mr. Donald R. Buchele
199 Fair Street
Berea, Ohio

Dear Mr. Buchele:

There is forwarded herewith for your personal files the original Grant of Letters Patent No. 2,679,183, which matured on 25 May 1954 from your patent application Serial No. 189,464, filed 10 October 1950, for a DIFFERENTIAL INTERFEROMETER. This original document which cannot be reproduced is being sent you by registered mail to ensure safe delivery. A soft copy of the patent will be sent to your co-inventor, Mr. Pierce B. Day.

It is requested that you complete and return direct to the Naval Photographic Center Patents Branch the enclosed receipt for the original Grant of Letters Patent.

Your cooperation in reporting your invention and your assistance in the prosecution of the patent application are appreciated.

Yours very truly,

Ira H. Abbott

J. W. Crowley
Associate Director for Research

Enclosures
Orig. Patent
Receipt form

cc: Mr. Pierce B. Day

***Letter 1:* Differential Interferometer Patent**
Copy of a letter to Donald Buchele, dated October 12, 1954, from the National Advisory Committee for Aeronautics, forwarding the Grant of Letters Patent No. 2,679,183 for the Differential Interferometer and identifying myself as a co-inventor.

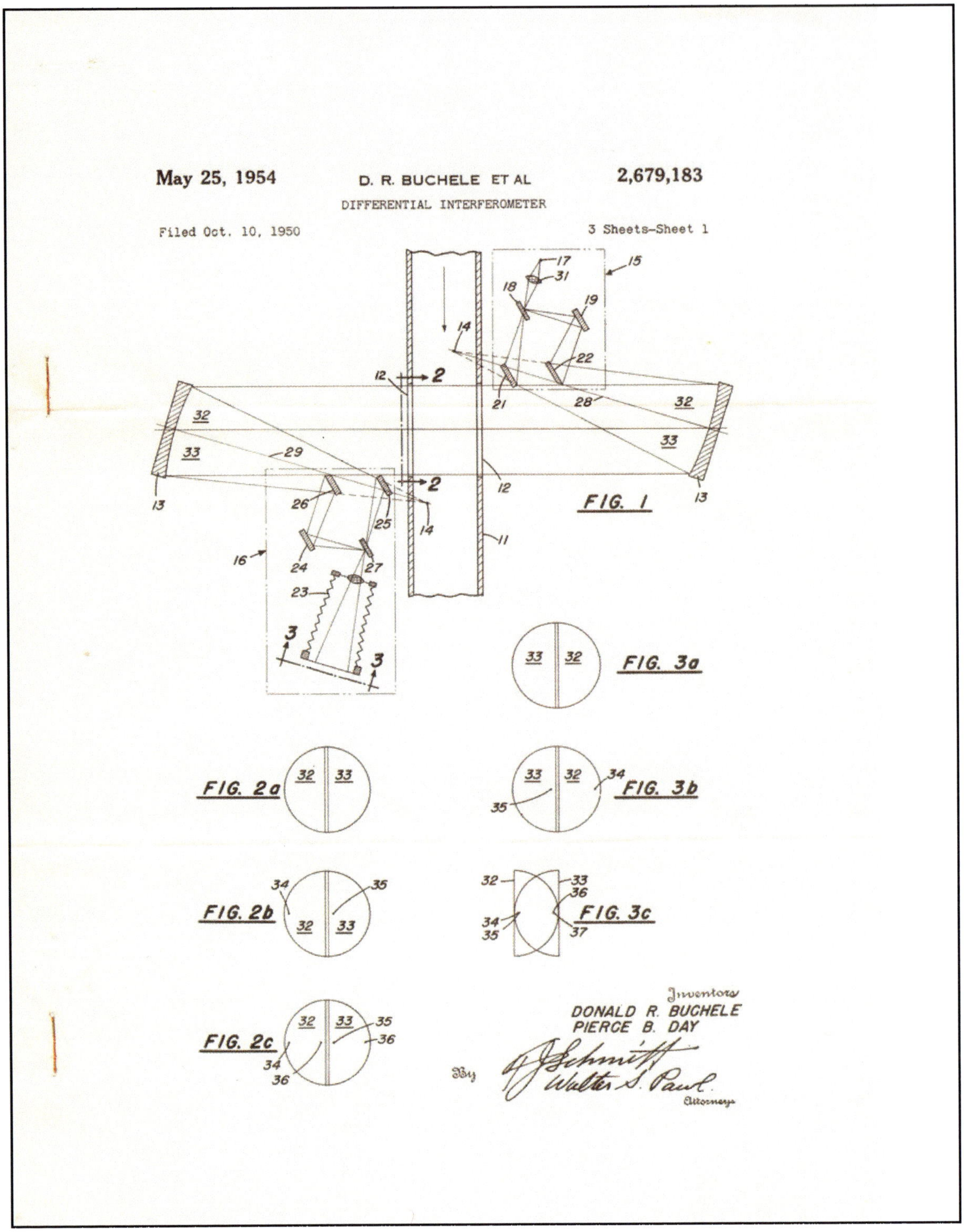

Figure 1: **Differential Interferometer Patent Art**
This is a page taken from the patent to illustrate the optics configuration. This was before the invention of the laser and unequal-path interferometers.

It was an upstairs flat on Rodenbeck Place — one bedroom, living room and kitchen- just off of Clinton Avenue south. Made a down payment on the place and headed back to pick up the car in Hamburg and my abandoned family in Fredonia. Since the Thruway hadn't been built yet, the trip to Rochester took close to four hours and we arrived late to a completely barren unfurnished apartment. However, the box springs and mattress I had ordered from Sears were in the hall at the foot of the steps. So at least we slept with a box spring and mattress between us and the floor. Jimmy's crib just fit into a large windowed closet off the living room.

3

A Graduate Student

To be back on familiar turf on the UofR campus was exciting enough but I was back home at the Institute of Optics with a sense of purpose and intellectually challenged by graduate courses in Math and Physics as well as Optics. I took Advanced Physical Optics, again with Parker Givens, and in Math an Introduction to Analysis. I barely made it through Electricity and Magnetism and had to scramble to catch up in a course in Modern Physics. I was intrigued by the fact that the earth is constantly bombarded from outer space by high-energy particles and we could actually record their tracks in a Wilson Cloud chamber. I also remember experiments in Spectroscopy — splitting white light and identifying the elements in the spectrum.

The laser was not invented until the 60's. There was no Laser Energetics Lab but at that time the Cyclotron was a pretty impressive facility. I was fortunate to have been there in the crucible of research leading to such dramatic developments both in Physics and Optics. My sole purpose, however, was to get through the course work and to get on with my own research project.

Some time in the spring of 1950 I launched into a project under the guidance of Gordon Milne to measure the brightness or radiant intensity of the nanosecond burst of light given off by exploding wires. Other investigators had estimated brightness 100 times the brightness of the sun but more accurate measurements were possible with the department's high-speed camera.

I had regular meetings with Gordon to discuss the methodology and the rather elaborate set-up, which took up an entire room. [I use the term methodology loosely to refer to the refinements of the experiment too complicated to get into. However, I can still see the set-up and feel the emotional impact that experience had on me]

To describe it briefly: a discharge circuit included two large 15kv oil-filled condensers which could be charged and discharged remotely using a metal bar on the end of a 5 foot long lucite rod. I was well aware that that rod was all that separated me from a potentially lethal jolt. A specially designed switch and resistances isolated the discharge circuit from the kv power supply. In addition there were two optical systems: one, a viewing system to align the images of the wire and the reference source on two slits. The other, to relay the slit images onto the film in the camera. To record the flash the department made available a high-speed strip camera. I would wrap strips of 16 mm film on the inside of a drum which was revved up to 12,600 rpm. The procedure then was to get the drum up to speed, and with my lucite rod, short circuit the discharge circuit, dumping energy like a bolt of lightning through these thin strands of wire. The wires were vaporized with a loud crack. (Again, my ear drums were subjected to a noise level well above EPA standards)

It took most of the summer to complete the set up so it was into the fall before I could start making exposures. By then I was working at Kodak and had to drive over to the campus evenings, go through this rather nerve wracking procedure and get back late at night. This went on for three years including the months it took to prepare the thesis. It was a monumental effort worthy of a doctorate. Looking back at those years I can only marvel at the tremendous energy, patience and determination it took but I had a goal and the help of a loving supportive wife. With her little Royal Portable typewriter, given to her in high school by her father, Sally typed every word of my 43 page dissertation- words that made absolutely no sense to her. Life was a three-ring circus with little Jimmy getting into things, me getting into my career in optics and finishing up a Masters program.

My First Patent

In the fall of 1950, I received two impressive letters with documents to sign from the Department of the Navy, Office of Naval Research. This was the second letter with a copy of the agreement I signed in October assigning the Government a non-exclusive, irrevocable, royalty-free, and non-transferable license.

November 9, 1950

Mr. Pierce B. Day
2 Rodenbeck Place
Rochester, New York

Dear Mr. Day:
Enclosed herewith is a letter addressed to you by the Office of Naval Research, Department of the Navy, dated November 6, 1950, together with an accepted copy of the License and Agreement in the patent application serial No, 189464, filed 10 October 1950, for the Differential Interferometer invented jointly by you and Mr. Donald R. Buchele of the Lewis Flight Propulsion Laboratory, Navy Case 11501, Aer-991.

Very truly yours,

J.W.Crowley,
Associate Director of Research

4

THE KODAK YEARS
1951 - 1986

How often do we see it coming, the life-changing events that shape our lives? And we seldom see or recognize the transforming influence, either direct or indirect. that people have in our lives. The telegram from Kingslake inviting me to join the lens design department at Kodak was certainly one of those events. After recovering from the initial shock, I had but one question — how soon?

Blessed with a beautiful wife and son and the prospect of a job at Kodak, I was grateful to accept my Masters degree diploma at the impressive commencement exercises in Fauver stadium.

Summer of 1950

I remember the royal treatment I got at Kodak Office. I was introduced to the head of personnel, Carl Lauterbach, who welcomed me into the Kodak family. I had the distinct impression that I could go to him with any problems I might have. I was much too confused, flattered perhaps by all the attention, to appreciate the benefits package. Kodak benefits were second to none in the industry and there were certain benefits of being a salaried employee. Speaking of salary, I would be making almost double what I made at NACA.

At the time I had no idea of the vastness of the Kodak complex in Rochester, let alone its world wide empire. But at least at that point in time, the three main building complexes were at Kodak Office/Camera Works, Hawk-Eye Works, and Kodak Park/The Research Labs. The camera design and engineering were all housed on Lake Avenue at Kodak Office, one of the original and most recognized buildings in the Kodak complex. Further north at Ridge Road The Research Labs literally straddled Lake Avenue with buildings on both sides and to the west of it was a huge area known as Kodak Park. Here were all the chemical and film manufacturing functions that go into producing that little yellow box. It was a city unto itself.

Hawk-Eye was both a very old and a relatively new building built during the war to accommodate the dramatic growth in orders for optics and optical equipment for the military. The old and the new were seamlessly connected floor to floor. With a new entrance on Driving Park, the new building extended along the Genesee River. The original building on St. Paul was the Photo Materials Company founded in 1891 by three former Kodak employees. Around the turn of the century Eastman bought out the company and acquired the building later naming it the Hawk-Eye Works after a very successful Hawk-Eye camera being marketed by the Blair Camera Company in Northboro, Mass. Eastman bought that one too and "relocated" it to the Hawk-Eye Works[1].

[1]Kingslake, Rudolf, "The Photographic Companies of Rochester, New York"
George Eastman House 1997

Photograph 5: **The Photo Materials Company**
The Photo Materials Company, ca. 1895, later acquired by the Blair Camera Company and George Eastman who named it the Hawk-Eye Works. Kingslake published this old photo of Hawk Eye in his book, The Photographic Manufacturing Companies of Rochester New York.

Photograph 6: **Hawk-Eye Works**
The Hawk-Eye Works and the Driving Park Bridge as seen from across the river. The new addition, built after the war, completely surrounds the old building to the south and west.

Kodak then introduced the Brownie Hawk-Eye and Baby Brownie both made at Hawk-Eye. As a boy in China my first camera was the Baby Brownie so it was strangely coincidental that I should have come half way around the world to work in the place where my camera was made.

The Lens Design Department

The Lens Design Department occupied the fourth floor of the old building and Kingslake's rather small office had windows facing St. Paul to the east and the School for the Deaf to the north. I can't remember that first day because I must have been very nervous meeting Kingslake and wondering what I was doing there or what was expected of me. I had always been in awe of him as a student and here I would be working under his constant scrutiny. He was the quintessential English gentleman with an aura of professor even in his capacity as a department head. Jinny Wolff, who I took to be his secretary, was a calming influence. Actually, as a mathematician she was much more than a secretary.

At the end of the war a number of Germany's most distinguished scientists immigrated to the U.S. Among them, two of their top lens designers who were now working for Kingslake. Otto Shade was tall and rather imposing. Spoke in a heavy German accent. My impression of him was that of a bespectacled Prussian general. The other German, Carl Tolle, was quite the opposite, mild-mannered and friendly. There were three people, other than Kingslake, actively designing lenses: Carl Tolle, Otto Shade and Fred Altman. Paul Stevens worked on projector optics and Don Feder was in the process of building a computer, as I learned later.

I ended up working for Carl Tolle, as an apprentice if you will, relieving him of the tedious job of ray tracing. Day after day I sat at a Marchant calculater tracing rays or calculating higher order aberrations. Back then lens design was an iterative process, based on the designer's experience and intuition, and labor intensive as I discovered. I believe the lens I worked on was a wide-angle version in the series of Ektar lenses for Kodak's SLR cameras. (Yes, at one time Kodak had one of the best SLR cameras on the market). Whether Kingslake intended for me to become a lens designer or not, I don't know. I can't remember having any burning ambition in that direction. By the end of the summer I had found friends in the engineering department and began thinking of making a move.

Department 35 — Engineering

Sometime in the fall of 1950 I moved down a floor to Department 35. Carl Novack, the Department Head, was a capable administrator but seemed to make no distinction between draftsman and engineer. Perhaps he thought of it as an essential learning experience for a brash know-it-all kid to have him work on the board for a while. I was placed at the far end of the room with all the tools of a draftsman — a board equipped with one of those wonderful tee-squares that rotate and lock in any position. I took mechanical drawing in the fall of '42, so I did have some experience. My first assignment was to create a perspective drawing of a rather complicated casting. I think I worked on it for months sitting next to a Charlie Shute, the department's lead designer whose work I soon came to admire.

SK-9647
84201-2

File

engineering file

EASTMAN KODAK COMPANY - - HAWK-EYE WORKS

OPTICAL DESIGN DEPARTMENT

Formula: 19-T-536

E.F.$_g$ = 10.13

Angular Field: 31½°

Wide-Angle Lens for 16mm Cameras

(Field further back than T-535)

Aperture: F/1.9

B.F. = 16.23

F.F. = -20.80

Lens	Glass	N_D	N_g	V
1 & 3	LBC-2	1.57250	1.58497	57.4
2	EK-310	1.74500	1.76538	46.4
4 & 5	BSC-2	1.51700	1.52690	64.5
6	BSC-1	1.51100	1.52096	63.5
7	EDF-4	1.75060	1.78644	27.7

Diam. of baffle: 11.20

Diaphr. Openings	
F/1.9	11.03
2.8	7.38
4	5.10
5.66	3.65
8	2.56
11.3	1.82
16	1.29
22.6	.91

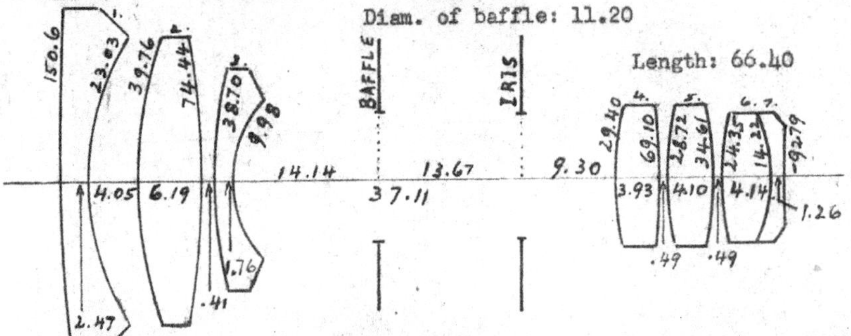

(Same as T-535)

Surface		Cl. Aper.	Trim	F/1.9 Axial Beam
1	34.20	33.80	35.50	5.33
2	29.80	28.66		5.30
3	28.30	28.04	29.50	5.76
4	28.30	26.70		5.77
5	21.80	21.54	23.00	5.74
6	16.60 bevel	16.28		5.56
7	13.30	13.02	14.50	13.02
8	13.20	12.94		12.94
9	13.00	12.73	14.50	12.73
10	12.30	12.01		12.01
11	11.30	11.02		11.02
12	—	10.65	13.00	9.54
13	11.00	10.65		8.78

Spherical Aberration		O.S.C.
F/1.9	+.049	+.00206
F/2.7	-.009	-.00040

Axial Color (g-D) = -.014

Lateral Color (g-D) 26° = .000
17.5° = -.006

Petzval Sum (100) = +.00182

Rim Rays:

17.5° A	17.5° ΔH'	26° A	26° ΔH'	31° A	31° ΔH'
-5.45	+.027	-10.38	+.017	-14.57	+.013
-5.90	+.007	-10.70	+.009	-14.80	+.011
-7.35	0	-11.92	0	-15.85	0
-8.97	-.009	-13.29	-.010	-16.97	-.012
-9.74	-.024	-13.95	-.030	-17.50	-.029

Field	Dist.	X'_s	X'_t
17.5°	-.066	-.054	-.014
26°	-.189	-.092	-.009
31°	-.240	-.067	+.066

Department Copy 7/17/57 msd

DESIGNER	K. T.	ASSIGN.	959
COMPUTER	K. T.		
DRAWING BY	K. T.	DATE	7-9-57

HE 850

***Formula 1:* The Lens Formula**
This is the lens formula for Kodak's 16mm movie camera, dated 7-9-57. Working for Karl Tolle in the Lens Design Department, I had the tedious job of tracing rays through one of these wide-angle lenses using a Marchant calculator.

EASTMAN KODAK COMPANY -- HAWK-EYE WORKS

Formula 25-M-52 (224K-2)
OPTICAL DESIGN DEPARTMENT
Speed F/2.5

Equiv. Focus: 179.80
LENS SPECIFICATIONS

Angular Field 26° (5"x5")
7" Aviation Lens

Back Focus: 123.13
Front Focus: 136.79

Lens	Glass	N_D	$N_{G'}$	V
1	DBC-3	1.61088	*1.62448	57.2
2	DBC-1	1.61100	1.62421	58.8
3	LF-1	1.57250	1.59011	42.5
4	DF-2	*1.61680	1.63909	36.6
5	EK-310	*1.74485	1.76575	46.4
6	EK-330	*1.75512	1.77572	47.2
7	LF-3	1.57500	1.59317	41.4

*Non Standard Indices

Diaph. Openings	
f/2.5	52.35
2.8	46.85
3.5	37.58
4.0	32.91
5.6	23.53
8	16.48
11	11.98
16	8.24

Length: 93.83

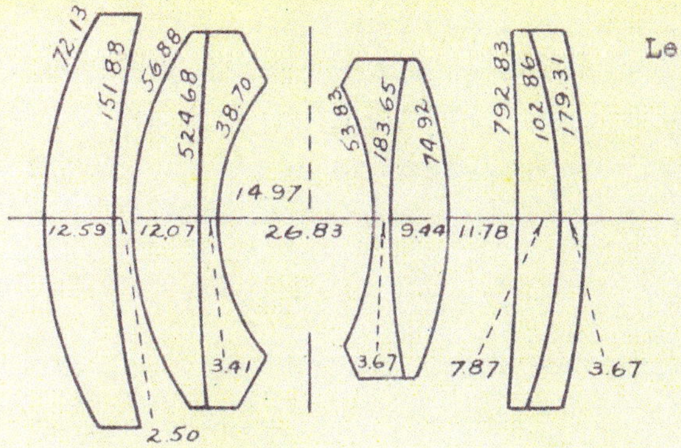

Surface	Cl. Aperture	Axial Beam
1	72.00	72.00
2	69.10	69.06
3	65.50	65.46
4		63.44
5	53.32	53.31
6	51.32	51.16
7		54.61
8	55.32	55.13
9	61.75	52.13
10		51.76
11	65.00	51.03

Spherical Aberration F No.	G'	D	Coma
2.5	+.02		+.14
3.5	-.52	-.52	+.07
Parax.	.00	+.07	

Lateral Color 18° (G'-D) +.002

Petzval Sum (100) = +.001611

Field	ΔY	ΔF"	ΔF'
18°	-.07	-.84	-1.02
24°	-.35	-.47	-1.05

Re-issued Apr. 3, 1951

DESIGNER G.A.	FILE NO. 12
COMPUTER	PAGES
DRAWING BY GT	DATE

***Formula 2:* Aviation Lens**

A 7–inch Aviation Lens for Aerial Reconnaissance, produced at Hawk-Eye during the war.

Photograph 7: **T-46 Stereo Rangefinder Prisms**
The T-46 Stereo-rangefinder prisms: on the left, a deflecting roof prism and on the right a deflecting Penta prism. The T-46 stereo rangefinders were installed in the Sherman tank.

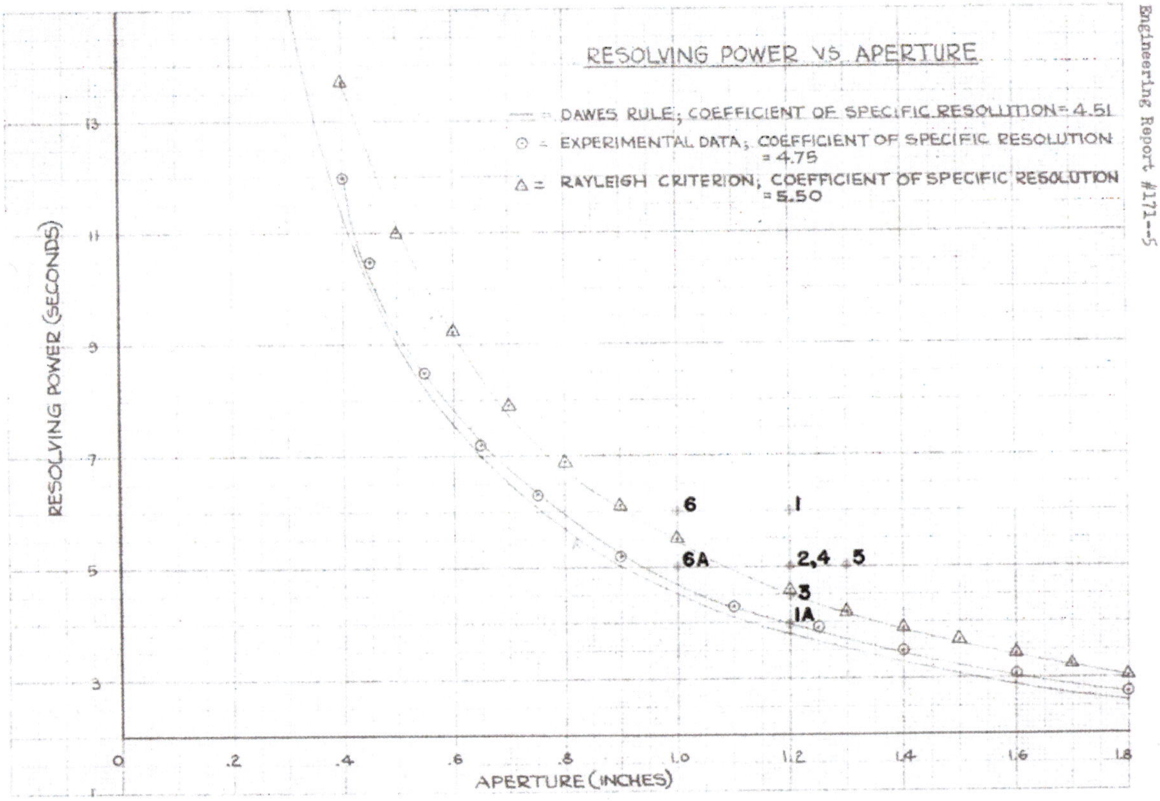

Figure 2: **Dawes Rule**
Curves of resolving power vs aperture for Dawes Rule, Experimental Data and the Rayleigh Criterion. The experimental data agreed more closely with Dawes Rule and provided a more accurate basis for tolerancing. Individual prism quality is indicated by the numbered data points.

Eventually my status was raised to that of an engineer although they really didn't have a job category for me. They had mechanical and electrical engineers but not optical. So I was, or became, the first Optical Engineer associated with the Engineering Department. Other Institute of Optics graduates who had preceded me at Hawk-Eye were essentially manufacturing engineers operating in different capacities. I should mention here that there were seven Institute of Optics graduates already at Hawk-Eye[2].

Aerial Camera Lenses

Although I was not directly involved, I remember one of the older mechanical engineers in the department, Charlie Lee, who had responsibility for the Ektar aerial camera lenses in production at Hawk-Eye. These were produced during the war and well into the fifties for aerial reconnaissance. I had thought that a 48 inch version of one of these lenses for the Air Force might have been used on the U-2 flights over Russia and fallen into Russian hands when Gary Powers was shot down in May 1960.

What struck me as strange about Charlie Lee was his apparent insecurity at a time when jobs at Kodak were as secure as it gets. He let it be known that if his job ever became superfluous, he kept a letter of resignation in his desk.

The T-46 Rangefinder

In retrospect, much could be said about the highly developed technology in the manufacture and testing of these aerial camera lenses. While I was not directly involved, I had a vested interest in the process of optics manufacturing and image quality criteria. I often stopped in on Mike Bernhardt, who was head of the optics test lab, to tap into his vast knowledge of lenses. Later on I would come to rely on his extensive collection of test data. The T-46 Rangefinder

Perhaps the first large government contract was the development of the T-46 rangefinder. It was carried out totally within Department 35 with Fred Wolff the program manager. The team was made up of a physicist, Charlie Bridge, two mechanical engineers and myself. I had been elevated because I knew optics. Fred was also a graduate of the Institute and as it so happened a fraternity brother, which was kept strictly between the two of us.

The stereo-rangefinder had been proven superior by the Germans in the historic naval battle in which the battleship Bismarck sank the pride of the English navy, the Hood. Of course I was not aware of this at the time but I suspect Kingslake's German lens designers were well aware of the rangefinders on the Bismarck. The optics for a rangefinder would normally be straightforward except that fitting it into a Sherman tank presented a unique problem. Since the tank commander stood to the left of center, the optical path to the right side of the tank had to have relay optics, which inverted the image.

This called for a roof prism to erect the image. On the right side a Penta prism was required. What really complicated matters was the fact that the optical axis angled downward and all the

[2]Neumer, Arthur E., BS, 1940; Bernhardt, Myron, BS, 1940; Rosborough, Robert S., BS, 1948; Harper, David C., BS, 1940; Schlauch, John E., BS, 1949; Stateler, Jack G., MS, 1949; Wolff, Fred J., BS, 1941; MS, 1953

prism surfaces tilted accordingly. It was Paul Stevens, I believe, who designed these prisms but Charlie Bridge and I had the fun of incorporating them into the system. I came up with the optical schematic.

As one might expect, mounting optics in a tank was another challenge. They had to withstand not only the vibration but the shock of a direct hit, requirements presumably provided in the contract. So new methods of mounting optics had to be devised and all the optics sub-assemblies shock tested. In the case of the two prisms, the most practical approach from a manufacturing standpoint was to bond them to a metal base, which then could be held in place with screws.

A small group of manufacturing engineers under Dr. McLeod had the formidable task of developing a bonding tape that was strong enough to grip the prisms and yet compliant enough to not crack them under the extreme temperature and shock environment in a tank. Needless to say a significant effort went into the development of this bonding tape, which could be produced in large sheets and cut to size. The acid test was to drop the prism assembly from about thirty inches onto an iron anvil. I don't know how many dummy prisms were "decapitated" on the guillotine before they had an unbreakable bond. The best fun of course, toward the end of the program, was to see how well we could range on a target. I have the distinct impression that we set up in Kingslake's office, since he had the only windows looking out to the east. As I remember we had a perfect "target"- the twin steeples of Saint Stanislaus Church on Hudson Ave. roughly 1.5 miles as the crow flies. No one thought of the irony of using a church to calibrate an instrument destined to be used in the next war.

We had thought World War II was the last war. Everyone was optimistic about the future. On December 22, 1952, we added a boy to the family, Robert Burton, through adoption. Then on February 25, 1954, just fourteen months later, we were blessed with the arrival of a daughter, Margaret Carroll.

Engineering Report #171

What really solidified my job description as an Optical Engineer was the responsibility to assign manufacturing tolerances on all the T-46 optical elements. This report, in fact, marks the beginning of a career dedicated to predicting image quality and the performance of optical systems. It was the first time I could combine basic science and research with the practical aspects of optical engineering. I was at last applying what I learned at the Institute of Optics.

The subject of the report was "The effect of aperture and axial astigmatism on the theoretical limit of resolution". Every student or graduate of the Institute is familiar with Dawes formula, which gives the fundamental relationship between resolving power, aperture and astigmatism measured in rings of irregularity:

aT = 4.51[square root sign] 2N

Where a is the minimum angle of resolution in seconds, T the aperture in inches, 4.51 a constant defined by Dawes Rule, and N the number of rings or fringes of irregularity as measured on an interferometer. Experimental results were qualified by the statement: "It is apparent that the formula applies to pure astigmatism and to produce it experimentally a device analogous to Coleman's astigmatism modifier was employed". The experiment consisted of three parts: a calibration of the astigmatism modifier, the measurement of resolving power vs pure astigmatism, and resolving power interferometer tests on fifty two T-46 prisms. The experimental results provided a basis for tolerancing as well as customer confidence in the manufacturing process.

The calibration of the astigmatism modifier is of particular interest because it presents actual measurements of resolving power vs aperture. The coefficient of specific resolution, 4.75, represents our attempt to duplicate Dawes' rule and the Dawes coefficient, 4.51[3]. The slight difference can be attributed to the contrast of the targets used. Resolution targets were reduced in contrast, in this case 2:1, to correspond more closely with what the tank commanders see.

Bombsights and Fire Control Sights

There were at least two contracts left over from the war years and they were important contributions. As the engineer (more or less) responsible for the T-46 optics, I had occasion to stop down a floor to Milt Yaekel's Optics Polishing Department, which occupied almost the entire second floor. It would be no exaggeration to say that it was the largest polishing room in the country. For me it was like a candy store to see the tremendous variety of glass in process. Polishing machines seemed to be running continuously on the high volume commercial optics. While it was not advertised, prisms and hemispherical domes were also being polished in volume on military contracts.

Some years later I learned that the double dove prisms and domes were being supplied to Farrand, a company in the New York area, for the Norden Bombsight and also the Hemisphere Sight for the B-45 Fire Control System. Fred Wolff reminded me just recently of the secret of the double doves in these systems. Because the two dove prisms (azimuth and elevation) are necessarily offset from the center of curvature of the dome, they would normally introduce unacceptable scanning errors. The solution was in the cement layer. In cementing the two halves of the dove a pre-determined wedge was built into the cement layer.

The B-52 Hemisphere Sight

About the time I arrived at Hawk-Eye in 1950 Kodak won a contract to design and produce the B-52 Fire Control System or Hemisphere Sight. It consisted of the head assembly and relay optics similar to the B-45 system but reconfigured for the B-52. When compact television cameras became available in the late '50's, these were incorporated in the B-52 system. It was a major redesign but the head assembly remained the same. This gave the tail gunner leg-room with the TV monitor located toward the front of the plane.

Comfort was not always a luxury for those who manned the guns in the World War II bombers. A recent autobiography details the hair-razing experiences of a World War II veteran, Chester Nothhard, who served as the turret gunner in the bubble on the under side of a B-24 Liberator.

[3]Wikipedia gives the Dawes' coefficient of specific resolution as 4.56.

Photograph 8: **The Norden Bombsight**
The Norden Bombsight, made by the Farrand Company, was used during the war on bombing runs over Europe. Kodak supplied 90 degree prisms as well as glass hemisphere domes.

Photograph 9: **The Hemisphere Sight**
The Hemisphere Sight for the B-25 Fire Control System. A rotating Dove prism provided scanning in elevation. The entire assembly rotated in azimuth.

A Periscope for the Republic F-105 Fighter Aircraft

Still a lowly engineer in the company hierarchy, I was not privy to the details of the contract with Republic except that it was to be a design study with delivery of a demonstration model. The project could now be placed in Department 35 rather than the Lens Design Department because it involved building a prototype and I suspect it was Kingslake who said I could do the optical design. I was flattered but had no idea how to approach the problem.

In high performance fighter aircraft such as the Republic F-105 the high angle of attack during landings made it difficult, if not impossible, for the pilot to see the runway. As I saw it, a periscope of that size had to be light-.weight and probably a dual system looking out through windows on the sides of the aircraft. The optics in each leg of the periscope would consist of a six inch diameter achromatic doublet, all plastic relay lenses and plastic field lenses where the two systems joined. The two six inch diameter exit pupils overlapped so that the pilot would have a contiguous but uninterrupted view of the outside world.

During the design phase I traced rays through the entire system as if it were a single lens, which was not an approach prescribed in Kingslake's Optical System Design. However, as any optics graduate knows there is only one way to orient two plano-spherical elements in a relay system and field lenses contribute little to image quality. So once I had laid out the basic configuration, the ray tracing merely confirmed the positions of image plane and exit pupil. All the plastic elements were ten inches in diameter except that the field lenses were slabbed off on three sides giving the pilot a rectangular "window" roughly 9 by 18. Strategically placed baffles were built into the housing.

I'll never forget the moment of truth: staring at the buildings across St. Paul through this incredible, bifurcated periscope. Hardly unbiased as its designer I thought the overall quality was excellent. I had it set up on a table in an area just before you entered the cafeteria so that everyone could take a look and pass judgment. I was indignant when some one christened it the "Iron Maiden".

As was often the case the customer expected delivery yesterday. I remember when the decision was made, just before the weekend, that Roger Coakley and I would drive down with it in his station wagon. We would deliver it on Monday because apparently a parallel study contract had been awarded the Institute of Optics and they had already delivered their version of a periscope. The designer, much to my surprise, was none other than Bob Hopkins himself. The two periscopes were set up side by side on the tarmac outside one of the hangers. Hopkins was not there at the time and in fact neither of us would hear the outcome of the "competition" until weeks later.

Before leaving they took us through the assembly area where we saw a mock-up of what I believe was the Grumman F-14A Tomcat. To see that fighter jet up close, with its huge air scoops on either side and swept back wings, was pretty impressive. There was no doubt in my mind that neither periscope would fly. Hoppy's all-glass periscope was way too heavy and my light-weight version suffered from color fringing around the edges. In the weeks that followed I went back to the drawing board, as they say, and came up with a much-improved design but by then no one was interested.

Photograph 10: **Test Fixture**
Test fixture used to measure power and astigmatism introduced by a shallow dome and prism combination when scanning in elevation. To compensate, ophthalmic lenses were inserted and later translated into a prescription for the toric hypotenuse side of the prism.

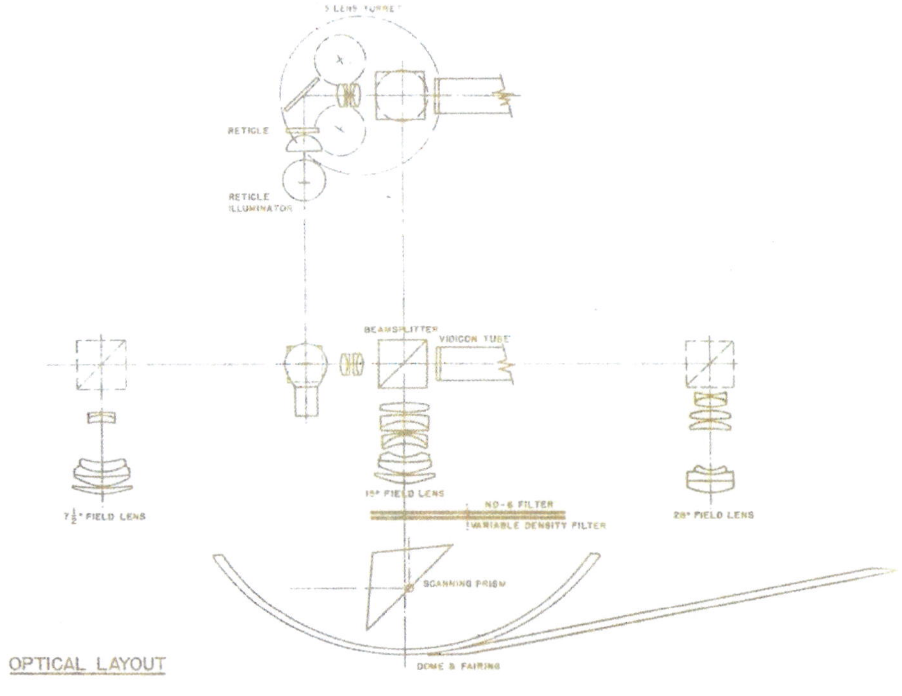

Figure 3: **North America TV Sight**
Optical layout of the North American TV Sight produced by Kodak for the Navy

Through the years I have recalled this experience with considerable satisfaction having competed with Robert Hopkins in an optical design competition that ended in a draw.

At Hawk-Eye things began to change fast. People began to disappear, including my boss, Fred Wolff, and highly sequestered programs were formed. I was transferred to Al Fultz's department and for months I was in limbo, feeling somewhat abandoned. My friends Charlie Bridge and Bob Uehlin had already left the company to work for Litten Industries in California and both wanted me to join them. In the fall I had a job offer from Litten and flew out with my wife to consider it. As is often the case, its not only about the money. My wife had a number of concerns about the school system and the high cost of living in the style they had become accustomed to. So I declined.

I had no idea what lay ahead for me at Kodak. On the home front, the family expanded once again with the birth of our second daughter, Sarah Jean, on August 24, 1960 - thus restoring the gender balance.

The North American TV Sight

In the late 50's the Navy came to Kodak with a requirement for a TV sight that would be mounted on the underbelly of its North American Fighter-bomber. The concept was to have a TV camera aimed straight down with a single prism to scan in elevation and presumably the camera and prism would rotate about a vertical axis. The problem was, however, they wanted to look through a glass bubble in the shape of a salad bowl. Aerodynamics dictated a shallow dome somewhat thicker at the edge for strength.

When I inherited the job it actually came with a fully polished dome. I was pretty much on my own to solve the problem so I chose what looked to be the easier path, which was experimental rather than an analytical approach. The latter seemed formidable. That may have been a mistake but at least I had in mind an experimental approach to a quick solution. I had a fixture built in the wood shop which provided for mounting the dome, a viewing telescope and a scanning prism. I also had a complete set of opthalmic lenses, which I could insert like an optician. Scanning from the middle of the dome to the edge I recorded, in diopters, the change in power and astigmatism. The problem then was how to compensate.

It seemed reasonable that with the appropriate curves on the prism hypotenuse it might be possible to compensate power as well as astigmatism. At least it was worth a try since there were really no other options.

The prism hypotenuse turned out to be a torric surface and I had to ask Milt Yaekel if he had anyone who could do it. If you knew Milt, he was never one to turn down a challenge and this was a good one especially considering the fact that these prisms would ultimately be required in quantity. He assigned the job to his top polisher, who I remember as having a very positive attitude like his boss. We got to be good friends because I would stop down every day to check on his progress. When he finally showed me, in the test glass, the right number of fringes in both directions, I gave my approval. He had done it.

Photograph 11: **General Eisenhower**
General Eisenhower addressing the troops on D-Day, June 6, 1944.

"You are about to embark upon a great crusade... good luck!
And let us all beseech the blessing of the almighty God
upon this great and noble undertaking."

-- Gen. Dwight D. Eisenhower
addressing U.S. troops before
the D-Day invasion, June 6, 1944

"General Eisenhower Addressing the D-Day Troops" has been made available to the World War II Veterans Committee courtesy of the National Archives.

WORLD WAR II VETERANS COMMITTEE

A PROJECT OF THE AMERICAN STUDIES CENTER
P.O. BOX 96543
WASHINGTON, D.C. 20090-6543

Figure 4: **General Eisenhower addressing the D-Day troops**
"General Eisenhower addressing the D-Day troops" was made available to the World War Veterans Committee courtesy of the National Archives.

Judgment day finally arrived and I was the defendant on trial for malfeasance. Allen Fultz was the judge and jury. What chance did I have for clemency? As head of the tool department, Fultz had, shall we say, an uncompromising reputation. His verdict was guilty as charged. The prism compensated beautifully over most of the scan, but in the most critical 10 percent, toward the edge of the dome, the quality of the image began to deteriorate.

In retrospect, I might have been able to improve the correction at the edge of the dome at the expense of the rest of the scan- using my fixture and the opthalmic lenses. But my time on the project ran out. I was transferred across town in the spring of 1957.

My invention report No. 89076 titled, "A Toric Surface for the Astigmatic Correction of a Shallow Dome and Prism Combination," was signed October 9th and forwarded to D.I.Mayne, the Patent Department, Kodak Office, October 30, 1956. It stated that the optical design study resulted in the best choice of radii to correct both the power and astigmatism introduced by the dome.

Two years later, long after I had left Hawk Eye on another assignment, someone came up with different radii for the prism. It was simply a different compromise between edge performance and the rest of the scan.

This precipitated what would have been a disastrous move on my part — a decision to leave Kodak.

5

Sputnik

I remember the headlines and reading with fascination the public response, bordering on panic, to the Russians having launched "a Sputnik", the first earth orbiting satellite. This was October 4, 1957 and one month later Sputnik II was launched with a dog, suggesting that perhaps humans could fly in space, but that would be some time in the future. The U.S. response came in December when the Navy attempted to launch its Vanguard satellite which malfunctioned and was destroyed by fire on the launch pad. Public confidence in the nations technical capabilities hit a new low. With successive failures of five intercontinental ballistic missile tests, Eisenhower addressed the nation saying:

> "there was rarely a day when I failed to give earnest study to reports of our progress and to estimates of Soviet capabilities."
>
> "intelligence sources revealed several Soviet technical accomplishments that for the first time put American soil at risk of a nuclear surprise attack, including: test of a hydrogen bomb in 1953. ... and hints of a ballistic missile program".

U2 intelligence gathering flights over the Soviet Union ended in May of 1960 when Gary Francis Powers was shot down, giving the highest priority to satellite reconnaissance programs.

Inside the Perimeter

In time my clearance went through and I took an oath that I would never disclose where I worked or what I did. It was a small concession, I thought, for at the time the prospect of working on what I knew to be a program to put optics in space captivated my imagination.

In October of 1958 I moved across town to a cold cement-block building on Lincoln Avenue, where I occupied a cubical with Don Smith[4], a mathematician working toward a doctorate at the UofR. This was the beginning of a fascinating 16 years in Kodak's Research and Engineering (R&E) Division. It never occurred to me that one day I might want to include these incredible, memorable years in a memoir.

In 1975 I was transferred to Copy Products where I had responsibility for the optics, and then returned to R&E for a year, 1985-1986, when I retired. Ironically, with that move I was eligible to become an ITT retiree when Kodak sold the Research and Engineering Division in 2004.

The urge to write about this phase of my career was primarily to recognize a debt this country owes to the Institute of Optics for its direct and indirect contributions to the U.S. reconnaissance programs. Secondly, the existence of the optics community at Kodak, for the most part housed at Hawk Eye, was a national resource in the 50's when it was called upon to deliver

[4]Smith, Donald Luke, Professor, University of Rochester

Photograph 12: **Engineers and Staff-1958**
Picture taken at Lincoln Plant at the start of the SAMOS program. (Author is far left.)

History of Research and Engineering Directors

Year	Directors
1956-1957	F. E. Tuttle
1957-1973	A. B. Simmons
1973-1975	C. P. Spoelhof
1975-1980	H.F. Hicks, Jr.
1980-1985	G. H. Conners
1985-1986	J. R. Robertson
1986-1992	G. H. Conners
1992-Present	T. C. Ritzsimmons

Figure 5: **History of Research and Engineering Directors**
C. P. Spoelhof later became Vice President in charge of the Kodak Apparatus Division.

the optics for the cameras that were installed in most of the satellite reconnaissance systems. It is impossible to do justice in recognizing contributions this relationship made to the overall success of these programs.

The Charles Stark Draper Award was a notable effort on the part of the National Academy of Engineers to give recognition to the network of contributors throughout industry. Don Schoessler, one of five recipients, received the award on behalf of Kodak. But what was the Kodak contribution specifically, in terms of the unprecedented technical challenges, notable achievements in the sciences and engineering disciplines? There is a chapter missing in the history of those early years of reconnaissance satellites, the extraordinary efforts of the Kodak team.

It was never a large organization and the optics community within Hawk-Eye was only a small part of it. It is to this relatively small community of optics workers, from the production and quality engineers to the polishers and assemblers, that I wish to pay tribute. They worked long hours, as did everyone, to achieve what seemed impossible at the outset. In my book these are the unsung heroes in the hayday of optics at Hawk-Eye.

History and the Rationale

> "In September 1961, an agreement between the CIA and the DOD established the highly classified NRO (National Reconnaissance Office) to consolidate ... all overhead strategic reconnaissance programs ... managed by civilian rather than uniformed Air Force personnel. The Undersecretary of the Air Force became the Director ...[5]"

The pros and cons for greater openness are fully documented in Richelson's definitive published work, "America's Secret Eyes in Space,[6]" the chapter titled "Still Secret After All These Years." I have resorted to extracting brief excerpts to provide perspective and a preamble to what follows.

> "The rationale for declassification of much information concerning the KEYHOLE program goes far beyond the fact that the justifications for absolute secrecy are flawed. One rationale for greater openness is based on a concern for history and the informed discussion of public policy. History may seem like a trivial concern to some, but it is the basis for understanding the past and acting in the future. **The history of the KEYHOLE programs is a history of one of the most significant military technological developments of this century and perhaps in all history.[7]"**

Those who are charged with preserving national security have not hesitated to prosecute individuals for disclosing classified material or even referring to the NRO and the nations reconnaissance satellite programs in public. The following experts from the public record illustrate the serious nature of perceived threats to national security.

[5]"The Air Force", published by The Air Force Historical Foundation, p. 230
[6] J.T. Richelson, "America's Secret Eyes In Space", The U.S. Keyhole Spy Satellite Program, 1990
[7] ibid, p. 265

> "In a February speech on the floor of Congress, [Congressman] Brown noted that the United States was currently operating only one KH-11 and that the "KH-12" would be launched on the second space shuttle mission once such missions resumed. Later that spring he compared the KH-8, KH-9 and KH-11 spacecraft and noted that the name NRO was officially classified although 'anyone can read about [it] in various unclassified articles, reports and books'[8]".

A letter to the chairman of the House Permanent Select Committee on Intelligence, on which Brown was a member, "...asked that Brown be instructed to restrict such comments to closed sessions of the Intelligence Committee". In the case of Samuel Loring Morison, an analyst for the Naval Intelligence Support Center (NISC), the CIA response was more severe. Morison had supplied a classified photo of the Russian Nikolayev Shipyard, showing a Soviet carrier under construction, to a trade magazine, *Jane's Defense Weekly*. He also leaked KH-11 photos of a massive 1984 explosion at Severomorsk, a Russian weapons storage depot. Morison was convicted and sentenced to two years in prison.

That was back in 1987. Now, 21 years later, much of the KH-7 imagery has been released and/or is available to the public through the American Freedom of Information Act, the justification being that image quality is comparable to currently available commercial satellite photography. Now the irony is that KH-11's capability to provide imagery globally is being "marketed" to foreign governments, albeit with restrictions on targets and resolution. And yet the history of how the U.S. arrived at this unique position has been systematically repressed.

Richelson makes the point, especially valid in my case, that secrecy has severely limited access to individuals involved in classified programs.

> "And those individuals are not immortal. Many of those involved in the early programs have passed away. By the year 2013, several more pioneers will be gone.[9]"

The Cold War (1955 - 1983)

The performance of our satellites in space, in terms of the size of objects resolved, was and still is the most guarded of secrets. Ironically, the secret was that initially the projected quality of ground resolution would not be good enough to identify missiles or missile sites. Eisenhower's 1955 "Open Skies Initiative" to exchange military installation blueprints with regular aerial photographs" was the most incredible proposal aimed directly at the soft spot in the closed society of communist Russia. Kruschev, of course, had to refuse. This was the beginning of the prolonged confrontation and economic boycott of Russia, the "Cold War."

Kruschchev's refusal legitimized the need for space based reconnaissance systems. From that time on thousands were sworn to secrecy — engulfed in the dark underworld of espionage — never to see the daylight, the freedom of open civil society.

[8] J.T. Richelson, America's Secret Eyes in Space, P. 258
[9] ibid, p.266

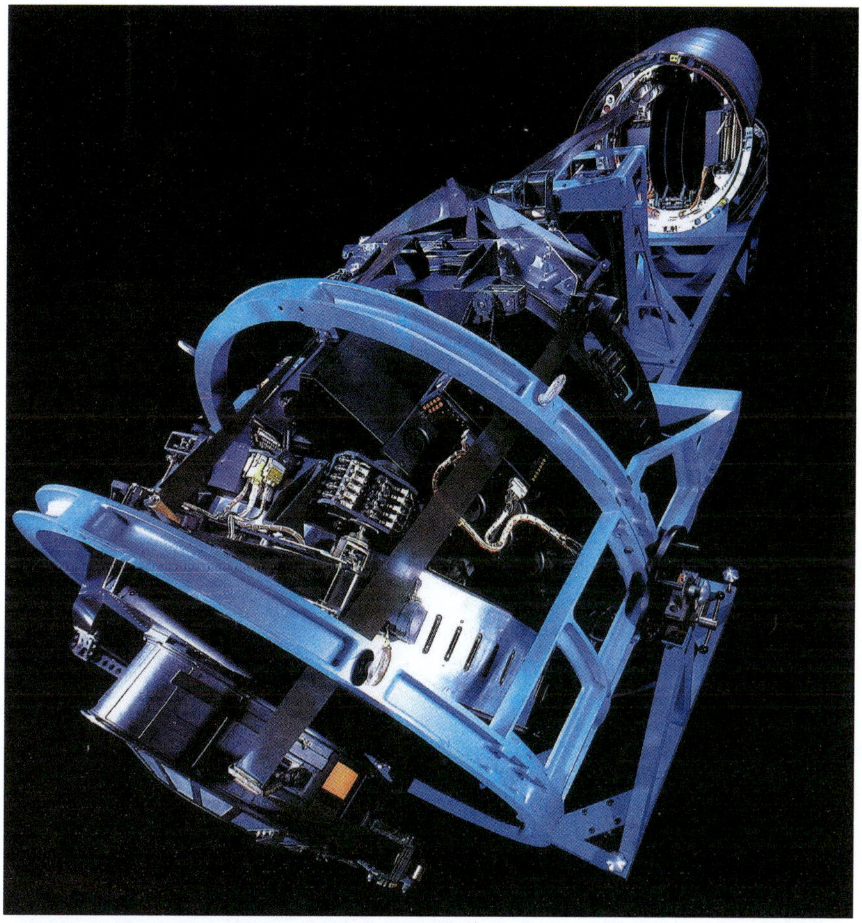

Photograph 13: **CORONA Reconnaissance Satellite.**
This photo and technical details were released when the program was declassified in 1995.

Photograph 14: **Midair Recovery of CORONA Satellite Film Capsule.**
On 19 August 1960, Air Force Captain Harold E. Mitchell piloting a C-119 "Flying Boxcar," and trailing a special nylon cable with retrieval hooks, made the first successful midair recovery of a CORONA satellite film capsule over the Pacific ocean on his third pass. (Jay Ashurst, Air Force Space Command)

The Question of Security

Anyone curious about the National Space Reconnaissance Programs, as I was, can down load information about the history of launches, contractors and technical details including resolution of objects on the ground. A full description of the CORONA satellite, the first in the series of reconnaissance satellites, was released in 1995 including a cross-section of the satellite, cameras, film path and recovery capsules- commonly referred to as "buckets.[10]" Of particular interest was the lens, a 24-inch focal length, f/5 Tessar as it was the first of a sequence of "lenses in space." In fact I was intrigued with the idea of tracing the sequence and evolution of these lenses, having witnessed the early stages of their development.

A *Smithsonian* publication[11] gives a brief history of the CORONA program, concluding with the statement: "The United States has continued operating photoreconnaissance satellites since CORONA ended, but they remain classified and little is known of them". On the contrary so much has now been published[12] about the satellite reconnaissance programs, including operational performance, that it begs the question whether continued classification is essential to national security. What is so fascinating about the data released on the internet, for example, is the fact that it gives quite detailed information on the optics. That is, ground resolution, focal length and in some cases, mirror diameter, from which the overall size can be inferred.

> "Richelson tells a fascinating story of technological accomplishment and bureaucratic infighting. Until an insider is willing and able to tell the complete history of America's photo reconnaissance satellites, this will be the definitive account."
>
> — Michael Krepon, President, The Henry L. Stimson Center.

This is one of three endorsements of Jeffrey Richelson's book, "*America's Secret Eyes in Space*" which of course stimulated this insider's interest in adding the Kodak story, more as a footnote to a history that may never be fully documented. I felt the need in some way to honor my colleagues, the engineers and technicians (the silent soldiers of the Cold War) who invested their working lives to make it happen.

The "Conspiracy" to Discredit SAMOS

Back to the internet free encyclopedia website, a 2007 entry on types of spy satellite, some interesting data is given for the Samos optics, that is, ground resolution and a range of focal lengths. It also mentions under "notes" that "Most flights used relay images; some film returns; probably cancelled due to poor-quality imagery". The program lasted only two years, 1960-1962, and was essentially written off as a failure. End of story.

[10]THE CORONA SATELLITE NRO Factsheet
CORONA PROGRAM PROFILE LockMart Factsheet 95-54 May, 1995

[11]Smithsonian Books, AFTER SPUTNIK 50 Years of the Space Age

[12]Deep Black, The Startling Truth Behind America's Top-secret Spy Satellites; William E, Burrows, 1986; "America's Secret Eyes in Space-The U.S. Keyhole Spy Satellite Program" Jeffrey T. Richelson, 1990; and "Secret Empire", Philip Taubman, 2003

As of 1963, the information leaked to the media consistently discredited the Air Force's SAMOS program, it "had not produced a single useful photograph.[13]"

> "... bloody thing never was workable[14]". But then the public record indicates that it did work - "Samos returned ... one set of photographs, with a resolution of 150 feet [30 meters was considered acceptable for mapping purposes].

Further, the inability to determine whether the photographs were of Russian or Chinese territory was clearly a problem of stability control[15] and not the operation of the SAMOS readout camera system. To be fair, it should be noted that the CIA's CORONA/Discoverer program was equally plagued by problems of stability and apparently "...produced little [useful imagery] in 1962".

This brings up the question of — to what extent is the public record slanted as a result of interagency rivalry and "bureaucratic infighting." Not that this is a concern to the general public, however, it reflects on the significant contribution of SAMOS to the success of America's space reconnaissance programs. Ironically the CIA's satellite program, "KH-11", as reported by Burrows:

> "... was the conceptual descendant of the Air Force's own SAMOS, a system the CIA had denigrated and tried to kill since its inception[16]" [!!!]

As one intimately involved with its development, I feel personally implicated in the crime and yet denied the freedom to testify to set the record straight. There is more to the story. Without some discussion of the concept, the technical details, and the probability of success, failure was perhaps too harsh an assessment. The fact that SAMOS was eminently successful in its reincarnation as the Lunar Orbiter clearly raises the question: "Exactly how did it fail and how did it perform as anticipated?"

In the interest of preserving the science, the history and, I might add, the reputation of that ground breaking program, I feel called upon to shed some light on the remarkable technology that originated with SAMOS, as well as, factors that may have contributed to its less than anticipated performance.

A distinction needs to be made between lens-design, a fully developed science, and the applied science, which was the development and production of the "Lenses in Space." A very significant lens design activity, therefore, can be considered open to discussion whereas the implementation of these designs, to function perfectly in space is an amazing story which has yet to be told.

Very early in the development of optics for satellite reconnaissance it was evident that a shift from refractive to reflective lens designs would be required as the lenses increased in size. The

[13] "America's Secret Eyes in Space", p68 - comment attributed to Richard M. Bissell, CIA

[14] "America's Secret Eye's in Space", p.68 - comment attributed to Robert Amory, CIA Deputy Director of Intelligence

[15] Note: It seems reasonable to suspect that the vehicle attitude control system may have been a factor, considering that the first successful Discoverer launch was not until August 1960.

[16] Deep Black, p.234

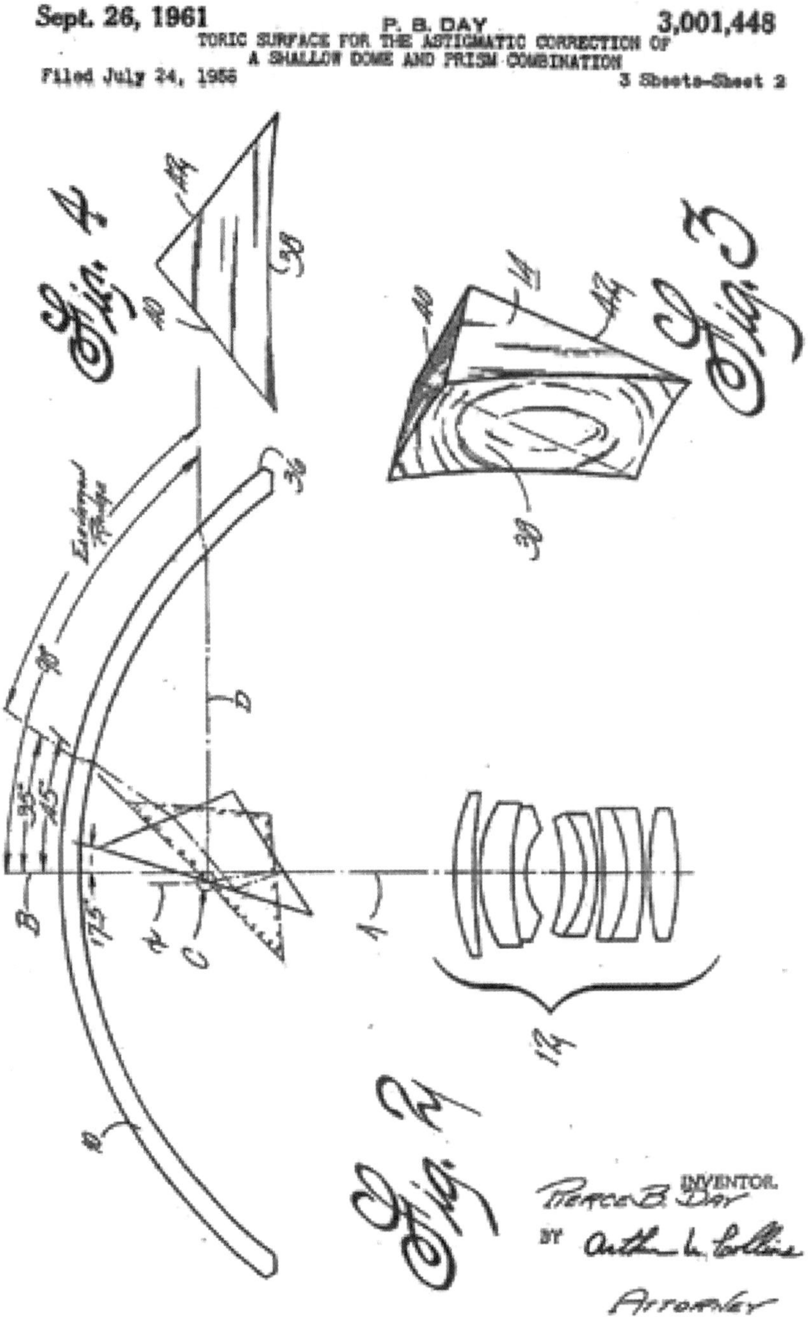

Figure 6: **Patent Illustration of Optics Configuration.**
Illustration of the optics configuration and a toric surface for the astigmatic correction of a shallow dome and prism combination, taken from the patent, which finally came through well after September 26, 1961.

Photograph 15: **Grumman F-14A Tomcat in Vertical Climb.**
A Grumman F-14A Tomcat intercepter in a vertical climb over El Centro, California. From this great U.S. Navy picture of the Tomcat, you can see what looks like a shallow dome and fairing directly under the nose.

first of these lens designs, developed for Samos, was a catadioptric design by Kingslake that combined refractive and reflective elements.

The irony is that some of these reflective mirror designs date back to Galileo and a Russian by the name of Maksutov. So, while there was a flurry of lens design activity for Samos and subsequent programs, the specific lens designs were essentially adaptations of designs long in the public domain. It was the scale of these lenses that was considered sensitive or confidential. The formulas were often, but not always, given for a focal length of 100 units.

The world in 1960 was in the grip of the cold war. The Corona photoreconnaissance satellite program was a "tightly guarded secret" until Lyndon Johnson announced to the world:

> "We've spent between $35 and $40 billion on space ...but if nothing else had come from that program except the knowledge that we get from our satellite photography, it would be worth ten times to us what the whole program has cost. Because tonight I know how many missiles the enemy has."

Thinking of Terminating my Kodak Employment

In the fall of 1957, I was transferred across town to a building where tanks were assembled during the war. The place was divided into separate secured work areas designated as "Black", some more black than others. From that time on I was not to speak to anyone, including family, about where I was or what I did.

Invention Report No. 89076 Resurfaces

Long after leaving Hawk Eye, I receive a copy of another report to the Head of the Patent Department, F.M.E. Holmes, dated March 26, 1958. The report is in answer to questions raised by Collins in connection with I.R. 89076, Docket 18334. My invention report, titled,

> "A Toric Surface for the Astigmatic Correction of a Shallow Dome and Prism Combination", was signed October 9, 1956. A year and a half had transpired.

I made the statement that the optical design study resulted in the best choice of radii to correct the astigmatism. So after I had left Hawk-Eye, someone came up with different radii for the prism. As for the patent, I never saw one.[17] The author of the report makes the following concluding remarks:

> "Since the invention report was submitted, extensive experimental tests have been made to improve resolution in the first engineering model of the NAA Sight. As a result of these tests, the curves on the prism hypotenuse are now weaker than those indicated in SK-9209, however, the principle of variable astigmatism compensation has applied successfully. This brings into question the statement in the invention report, "The optical design study resulted in the best choice of radii."

[17] At least not until April 2008 when I called the Kodak Patent Office and Beth Reece ran a search

Copies of this report went to J.L.Boon, Dr.K.G.Macleish and F.J.Wolff.

What bothered me about this report was the fact that the author seems to have taken credit for the concept as well as the "principle of variable astigmatism compensation." I was never consulted and I felt my reputation had been compromised.

In retrospect, there was no feedback with respect to the success of the program, but of course, I had "no need to know." The title "North American TV Sight" was all I needed to know, but I always wondered about the plane and how many of those incredible prisms were made. Sixty years later I acquired a magnificent volume, *U.S. Naval Aviation,*[18] in which I was able to identify the plane as the North American A3J (A-5) nuclear bomber — later the RA-5C Vigilante reconnaissance jet. The Grumman F-14 Tomcat, took over the reconnaissance missions, replacing the RA-5C's in 1979, and from this great picture of the Tomcat in a vertical climb over El Centro, California, looks on close examination to be equipped with a shallow dome and fairing directly under the nose.

A Job Offer from Perkin Elmer

After two financially devastating experiences with disreputable builders, we were finally able to build a small ranch in Perinton. It was completed in the summer of 1957 and we moved our family of three (two boys and one girl) from Regent Street to Hunters Run that Fall. The following year I worked on SAMOS — primarily with the optics.

In response to a rather attractive job offer from Perkin Elmer, my wife and I drove down to Norwalk for the interview followed by a tour of the housing market. Despite the significantly higher cost of housing, we came home and put the house we had just settled into on the market. It was that close.

I had gone to work the following Monday fully expecting to announce my resignation. I'm not sure if I spoke to Don Smith with whom I shared an office, but I do remember a conversation with Ken Macleish, apparently my immediate superior in the chain of command. He patiently heard me out and then proceeded to convince me that this would not be a particularly good move, that I had a future at Kodak. So for the second time I turned down a lucrative offer, end of story except that evidently Macleish had also been approached. Not more than a month or so later I learned that Macleish accepted a vice presidency at Perkin Elmer.

In retrospect two things played on my mind. First, I'm now convinced he did me a great favor and second, the irony was that hiring a young optical engineer instead of a VP might have averted the Hubble catastrophe.

[18]Naval Aviation Foundation, 2005 edition, p.103

Program Designations

Three distinct Samos programs are identified on the Internet: "SAMOS Projects 101A (E2) and 101B (E5) for, respectively, readout high magnification visual surveillance and film recovery high magnification visual reconnaissance; SAMOS Program 201(E6), for film recovery high-magnification visual reconnaissance. SAMOS was terminated (Projects 101 in 1961 and Program 201 in 1962), as CORONA achieved increasing success. Yet throughout its early years, Samos was reported extensively as the key spy satellite that keeps this Nation informed of vital military installations and build-ups behind the Iron Curtain.[19]"

[19]Spy satellite - Wikipedia, the free encyclopedia - The Corona Satellite NRO Factsheet

6

Samos

Project 101A (E2)

The early history of failures of the Discoverer (CORONA) program, which relied on the recovery of (unprocessed) film in buckets, placed unprecedented pressure on the Kodak program known as SAMOS. The proposal was to process film in orbit "down-loading" imagery via a telemetry link. This was a monumental challenge for Kodak both from the standpoint of organization and technology.

So how do you process film in space — an environment unfriendly to liquids? Evidently some work had been done at the Research Labs on a dry process called Bimat, but would it work reliably in the harsh environment of space? The camera with built-in processing would be state-of-the-art and in retrospect a technical work of art. I remember the person responsible for the processor, John Meyers, as being a brilliant young engineer with a contagious exuberance. The development of the processor/camera subsystem was crucial to the success of SAMOS and deserves recognition as an important innovative contribution.

In addition to the processor/camera subsystem, lenses had to be designed as well as the readout system and the telemetry link. Briefly, following exposure and development, the film is stored until the satellite passes over a ground station, which may be a number of orbits later. The readout subsystem is activated, the imagery scanned and transmitted to the ground station where electrical signals are converted to imagery on strips of film again. Obviously the timing and the sequencing of events were critical.

The telemetry link was analogous to flying a radio controlled model airplane but complicated for reasons of security. Remember, this was before the advent of digital technology and the Internet, so that transmissions were analogue. Ground stations had to be located strategically around the world to provide continuous satellite contact. When SAMOS was first proposed there were a number of technical unknowns not the least of which was the effect of the readout system, plus the ground reconstruction of imagery, on image quality.

The Readout System

There was nothing new about flying spot scanners. They had been used at Kodak Park to detect flaws during the manufacture of film and sensitized papers. These scanners used a CBS line scan tube, a lens that projected a spot onto the film and a photomultiplier to pick up light transmitted through the film.

One of my first assignments on SAMOS was to evaluate lenses for the readout system or more properly to survey the state-of-the-art. The rather interesting chart, Figure 7, is the evidence-preserved almost half a century. All of these lens formulas, with the exception of the Goertz Artar, were from Kingslake's file. Test data came for the most part from Mike Bernhardt's files. Two of the lenses, C and J, were presumably used in the ground reconstruction unit. The lens

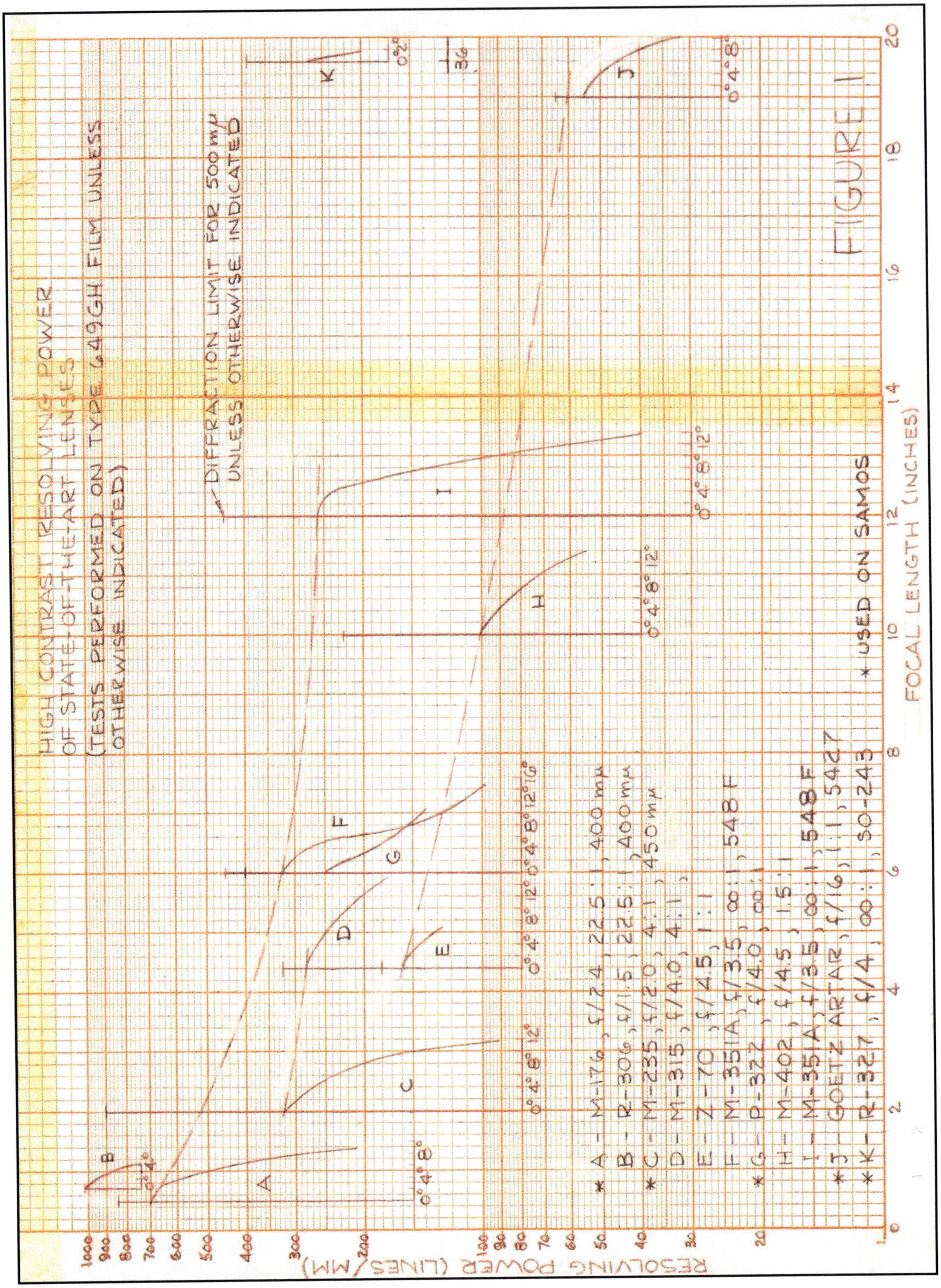

Figure 7: **SAMOS Lens Data**
An interesting chart, dating back to the early 1960's. It summarizes the resolving power vs focal length data for the actual or proposed SAMOS lens.

used to scan the film in orbit was A, the M-176 with a relative aperture of f/2.4. Comparing the resolving power of the two readout lenses, A and B, you might wonder why B, the R-306, was not used — clearly a better choice in retrospect.

Unfortunately, neither the formula[20] nor the formula sample survived — only a gem of an idea.[21] The R-306 was a miniature Maksutov telescope with a relative aperture of f/1.5! I remember how impressed I was with the simplicity of the design, just three elements: a small spherical mirror, a simple corrector plate and a 45 degree mirror cemented to the inside surface. The axial image is fully corrected for spherical aberration so that at 22.5:1 and in the green monochromatic light, 400mu, of the flying spot scanner it was diffraction limited. A formula sample was actually made and tested in Mike Bernhardt's lab. I still have the test report as evidence of the image quality attainable with a diffraction limited lens and this remarkable Type 649GH film: 1000 lines/mm falling off to 700 lines/mm off-axis. By comparison, the M-176 resolution fell from 700 lines/mm to 200 lines/mm off-axis.

However, the quality of the scanned image was limited by the spot size of the line-scan CRT tube and the decay time of the phosphor — not the lens. The spot size at the film was on the order of 5 microns (200 l/mm) in width and perhaps 7.5 microns in the scan direction. The flying spot had a tail like a comet.

The Lenses

There were two lenses used on SAMOS — a 6-inch and a 36-inch.

> "The E-2 lens was a 6-inch Petzval lens designed by the German designer Willy Schade. He was an expert in this type of lens, which was also used in slide projectors. It had one design feature that limited its field of view; field curvature. The two sets of doublets (both positive) added up to a slightly curved image. However, the use of a curved field flattener plate corrected this problem.[22]"

The other lens was the R-327 designed by Kingslake.[23] It had a 36-inch focal length operating at f/4. The "R" in the formula indicates that it was a catadioptric lens system consisting of both reflective and refractive elements.

This particular configuration is known as a Maksutov[24] with a primary spherical mirror and a single or double corrector plate in front. A small rectangular flat mirror is suspended in the converging beam to bring the image plane out to the side. In this design the field flattener is an achromatic doublet.

[20]Although I chose a career as an optical engineer rather than a lens designer, I can claim to have designed a lens, the "mini-Maksutov" R-306. Tragically, Kingslake's entire file of lens formulas was destroyed when Bill Fujimura died. Bill succeeded Bill Price as head of the Lens Design Department.

[21] There was no possibility of filing an invention report

[22] Contributed by R.E.Keim

[23] See the Institute of Optics 75th Anniversary Essays page 97

[24]Comment by R.E.Keim: "The formula is referred to as a Maksutov, but credit for this configuration should be given to Bouwers in the Netherlands who simultaneously developed the design but, because of German occupation, was unable to publish his findings until the end of WW2. It should be called a "Bouwers-Maksutov" but people shorten the name to Maksutov."

EASTMAN KODAK COMPANY - - HAWK-EYE WORKS

OPTICAL DESIGN DEPARTMENT

Formula: R-327

E.F.$_D$ = 906.4 (35.68")

Ang. Field: 1.6° (? inch diag.)

36-inch Catadioptric System for Lincoln Plant

(.500 to .700 μ)

Like R-326, with lateral color correction.

Aperture: F/4

B.F.$_D$ = 20.04

Lens	Glass	N_C	N_D	N_F	V
1	BSC-2	1.51461	1.5170	1.52262	64.5
3	*Yellow Filter	1.51736	1.5200	1.52629	58.2
4	DBC-2	1.61370	1.6170	1.62493	54.9
5	DF-2	1.61218	1.6170	1.62904	36.6

(* assumed values)

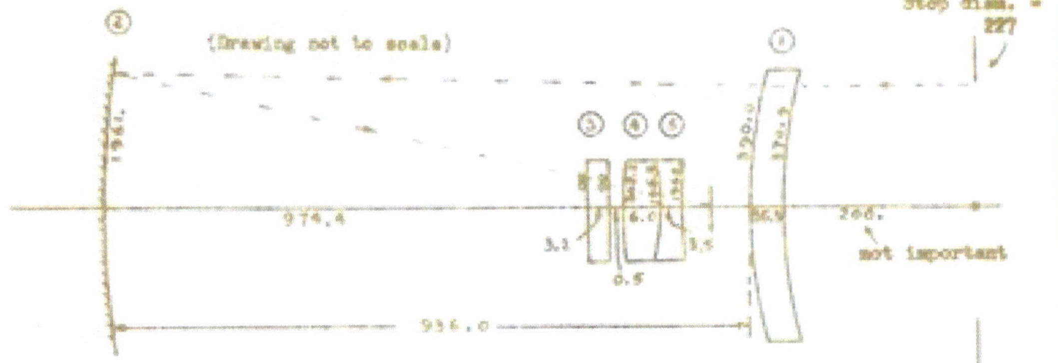

Surface	Cl. Aper.	Axial Beam
1	237.3	226.4
2	246.5	234.0
3 (mirror)	304.0	239.9
4	60.7	7.2
5	60.2	6.7
6	59.7	6.5
7	59.4	5.6
8	57.8	5.0

Spherical Aberration

F-no.	C	D	F	O.S.C.
4	.025	.021	.010	-.00001
5.6	-.013	-.013	-.014	.00002
Parax.	-.003	0	.008	—

Petzval Sum (100) = 000

Field	Dist.	X's	X't	Lateral Col. (F-C)
1.6°	.029	-.004	-.007	-.0007
1°	.007	-.002	-.002	-.0017

Rim Rays:

1.6° A	1.6° ΔH'	1° A	1° ΔH'
119.2	.0007	117.2	.0014
86.9	-.0008	84.9	-.0010
6.0	0	6.0	0
-74.9	.0025	-76.9	.0016
-107.2	-.0037	-109.2	-.0037

DATE 12/9/58

Formula 3: **Kingslake Formula Sheet**

This is the formula sheet, initialed by Kingslake, for a "36-inch Catadioptric System for Lincoln Plant". Although the SAMOS program and its location at Lincoln Plant were classified, lens formulas were generally unclassified. The date, 12-9-58, coincides with the start up of the SAMOS program.

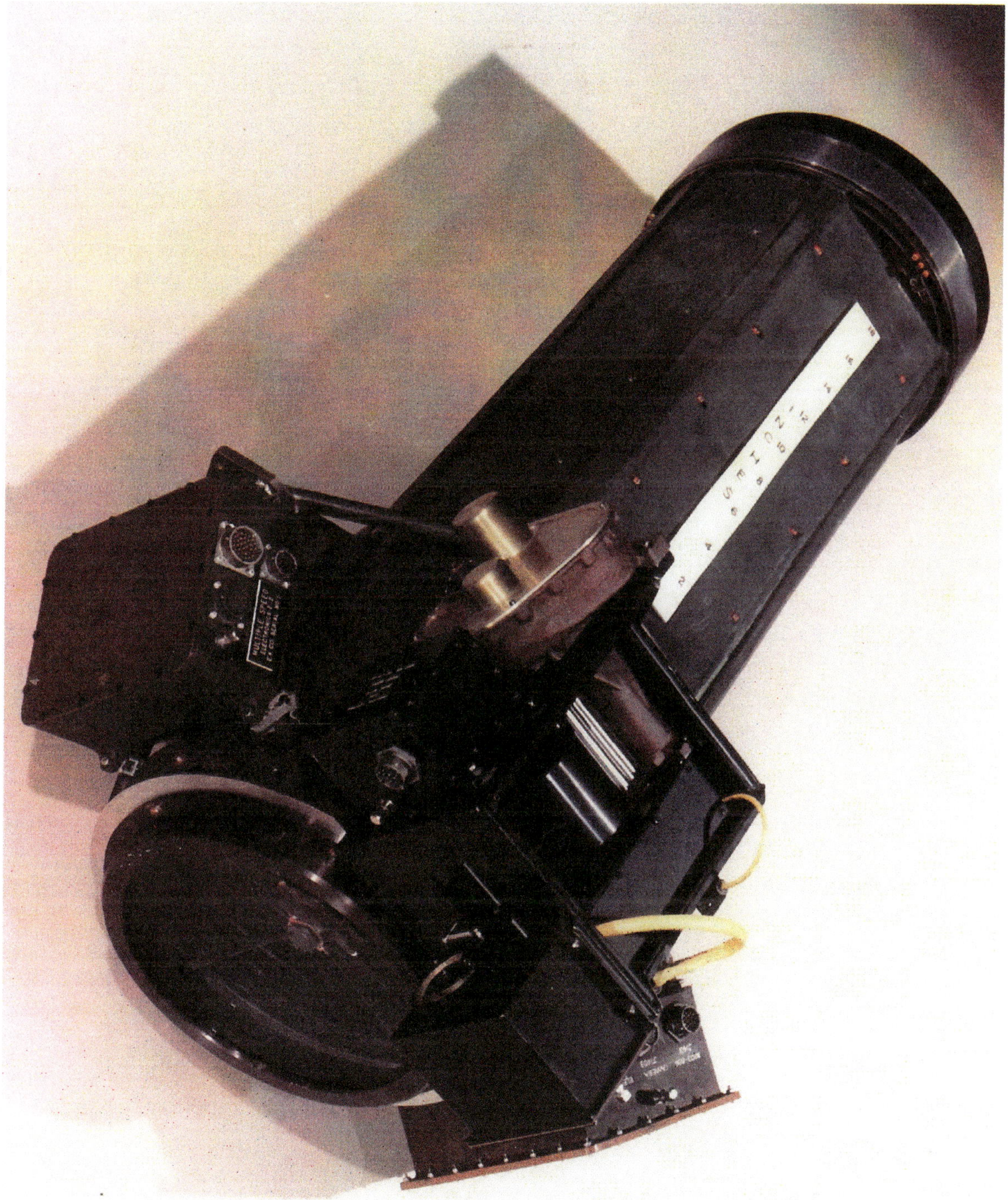

***Photograph 16:* 36-Inch Lens Assembly**
The 36-inch lens assembly including film drive and electronics. It was used on Program 201 (E-6), which was designated for "film recovery high magnification visual reconnaissance."

Photograph 17: **The Lunar Orbiter**
The Lunar Orbiter was the direct descendent of SAMOS. The film readout and telemetry worked perfectly while circling the moon.

Kingslake's R-327 Formula

It was common practice at Hawk-Eye, after the Lens Design Department issued a formula, to have a formula sample built and lens bench tests performed to verify the design. Referring to the chart, Figure 7, the test data shown for K is a plot of the single-pass resolving power for the 36-inch formula sample or prototype lens. The film, SO 243[25], is an early version of the film used on all of the subsequent film-based reconnaissance systems.

High-contrast resolving power, ranged from 275 lines/mm on axis to 200 lines/mm 2 degrees off-axis.[26] The diffraction limit shown at slightly under 400 lines/mm is based on the Rayleigh Criterion converting the angular resolution to lines/mm at the image plane. Taking into account the resolution limit of the film at 500 lines/mm, the perfect diffraction limited lens would have a resolution limit of about 330 lines/mm. So at 275 lines/mm the optical quality factor or OQF for this lens was 83 percent.[27]

This was remarkable considering the optics manufacturing and testing capabilities prevalent at that time. (Test plating the 9-inch diameter mirror and corrector element surfaces to 1/8 wave was state-of-the-art. The fully assembled lens would have been evaluated on a lens bench, essentially to check symmetry of the Airy Disc star image and again on a collimator.[28]) Laser interferometers were still not used for testing lenses.

With a lens/film resolution of 275 lines/mm, Kingslake's 36-inch lens could have produced ground resolution of 2 feet[29] under ideal conditions.

If the Wikipedia reported ground resolution of 1.5m or 5 feet was in fact achieved, then the difference could be attributed to a variety of image quality losses in operating a camera in space such as maintaining optimum focus, smear due to satellite instability or IMC (image motion compensation) error and inability to acquire resolution targets exactly at nadir. Its important to note that all these lens formulas exhibit a fall-off in quality off-axis and this is particularly true of Kingslake's 36-inch lens.

I was always intrigued by the fact that this type of lens was introduced first by a Russian named Maksutov and the fact that a Russian contributed the basic configuration of the optics in our "spy" satellites.

[25]The resolving power of SO-243 film was 500 lines/mm

[26] It was indeed fortuitous that an actual tri-bar measurement on this lens was preserved,

[27] A lens with an Optical Quality Factor higher than 70 percent can be considered diffraction limited.

[28]Descriptive material provided by R.E.Keim: "The 36-inch lens was tested on a 240-inch collimator which was designed to provide a full field-of-view test. Its primary was aspheric and a window, also slightly aspheric, was used to seal the optical path. The target array consisted of resolution patterns, aerial scenes and test patterns to determine film velocity."

[29]To convert from lines/mm to feet on the ground, convert the reciprocal of lines/mm to feet in the film plane and multiply by the scale factor (altitude in feet divided by focal length in feet).

Photograph 18: **Lunar Crater "Copernicus"**
This NASA image shows the lunar crater Copernicus, which is 60 miles in diameter and 2 miles deep. When it was taken by Lunar Orbiter II, it was hailed as "the picture of the century". The five successful Lunar Orbiter probes of 1966-67 were intended to map the lunar surface in preparation for the Apollo landings.

The Slit Aperture Plate and Flare

Keim reminded me of two design features that were critical to successfully implementing the R-327 formula:

> "A small, but significant, part of the formula was the slit aperture plate. Because the system operated as a strip camera, the shutter was a focal-plane device located near the film plane. The exposure time was equal to the slit width divided by the film velocity. However, as the film initially progressed behind the slit, a point on the film would first sense light from only the edge of the lens aperture. Further on, the point would be in the entire illuminating beam, then progress to the other side of the slit. As a result, part of the total time the point moves through the exposing region it would not receive the full cone of light from the aperture. This phenomenon produces a less than perfect image-forming beam due to diffraction effects. To achieve the maximum possible exposing profile the slit[30] must be as close as possible to the film. In this case the space between the slit and the film was only .004in. In this arrangement, a point on the film would sense the fully exposing beam at least 70% to 80% of the time.
>
> The plate was also a slightly concave cylinder on the side toward the lens. This curvature was designed to provide a fully flat image plane.
>
> Because of the folded configuration of the Maksutov lens (Its folding mirror and doublet corrector are very close to the aperture, which was the front of the meniscus lens.) it is susceptible to flare from the broad field of surrounding light. Consequently, the lens barrel was purposely painted gloss black internally and baffles were placed at locations which thereby caused all internal reflections to undergo numerous "bounces", each time reducing them in light intensity. A photo test showed that the veiling glare from external light would be less than 1%."

[30] Comment by R.E.Keim : "After trying other techniques, the slit plate was successfully made by the use of a diffraction grating ruling engine at Bausch & Lomb to scribe the openings on an evaporated aluminum plate."

7

Lunar Orbiter

It's strange that I never made the connection between SAMOS (1960-1962) and the Lunar Orbiter missions in 1966-1967, so complete was the separation between my clandestine work environment and the world I could read about in the papers. I was never directly involved in the assembly of the SAMOS hardware and totally isolated from Kodak's Lunar Orbiter operations, being a so-called "white" program. For that reason the lenses were procured from Pacific Optical. One was an 80 millimeter wide angle lens to provide moderate resolution coverage; the other a 24 inch focal length lens for high-resolution photographs.

Bob Keim[31] tells an interesting story about how the lens had to be slimmed down. An off-the-shelf Pacific Optical lens, a Petzval type formula, was chosen but would have added 10 pounds to the weight of the payload- completely unacceptable. John Villa, one of Kodak's lens designers, recommended a redesign with thinner elements but was told they wouldn't change radii of the concave surfaces. They didn't want to remake test glasses!

Some years later, on a trip to Washington with the family, we visited the Smithsonian and I was able to point to the Lunar Orbiter on exhibit there and say "We made that at Kodak." I must admit to having felt the pride of ownership, being able to identify myself with an object from space- a satellite.

> "The Lunar Orbiter, a robotic spacecraft equipped with a powerful camera, mapped the lunar surface prior to the first Apollo missions, paving the way for human landings. Five of these spacecraft flew to the Moon, entered orbit, and scouted the terrain to identify potential landing sites. ...Lunar Orbiter provided the most useful and detailed data. This data allowed NASA mission planners to select a safe and scientifically interesting landing site for each Apollo mission.[32]"

Reading further in the text that accompanied this superb picture of the Lunar Orbiter, I found the implied but obscure link with SAMOS: "Approved in 1960, the Lunar Orbiter program originally had no connection to Apollo"[!] That was because the "original plan" was the SAMOS program, the success of which was never publically acknowledged. In fact, the following excerpt from the current Wikipedia website essentially dismisses SAMOS as a failure.

> "Most flights used radio to relay images; some film returns; probably cancelled due to poor-quality imagery."

The data given for ground resolution, 30 meters, would justify that assessment except that image quality should not have been the issue so much as the fact that near real-time reconnaissance was successfully demonstrated. Further, as part of the 201 (E-6) Program, "high magnification visual reconnaissance," the 36-inch lens produced 1.5 meter or 5 foot ground resolution.

[31] Robert E., M.S., 1957, Institute of Optics

[32] After Sputnik, 50 Years of the Space Age, Smithsonian Books 2007

***Photograph 19:* Lunar Surface Mosaic at NASA Headquaters**
A 30 square foot mosaic of the lunar surface spread out on the floor at NASA headquarters. It was assembled by the U.S. Army map service in 1968 and is made from 127 photographs taken by Orbiter 4 in 1947.

This was with the diffraction limited 36-inch lens that had the potential of 2 foot ground resolution!

It was the successful recovery of film on the Corona program and the decision to move in that direction that prompted the cancellation of SAMOS.

I was left out of the loop on the transfer of SAMOS technology to the Lunar Orbiter program. However, when I was still at Lincoln Plant in early 1960, I remember seeing a layout of reconstructed imagery from SAMOS — the engineer in charge invited me up to have a look. Each 17-inch long strip of 35mm film represented a scan of the original 70mm film. When all these strips were laid out side-by-side on the floor, the composite image covered an area roughly 15x15 feet. I was impressed that he had to align each individual film strip. The Lunar surface image pictured on the opposite page could well have been taken from this reconstruction.

From the picture of the 30 square foot mosaic taken at NASA, assembled from Orbiter 4, edges of the 35mm film strips are clearly visible as well as the blocks of imagery that come out of the reconstruction printer then pieced together.

The Reconstruction Printer

As incredible as this "picture of the century" is, the technology that made it possible, dating back to the development of Samos, 1958-1960, would stagger the imagination of today's rocket scientist. This is how the picture "down-loaded" from space was reconstructed:

> "The air-ground telemetry of the scanned images was used to produce an initial ground record on 35mm film. Each scanning pass on the original 70mm film produced a 1-inch x 17-inch image on the 35mm film. However, because the scans were in an alternating across-then-back motion, the images on the 35mm ground film also were alternately head-to-tail-to-head, etc. The next operation was to reconstruct and assemble these individual images in their proper orientation."

This was done by the use of an optical array working at unit (1:1) magnification. It consists of two optical paths, independent of each other but both viewing the original [back-lighted] 35mm film and focused on the final reconstructed film plane. The first was simply a straight through film-to-lens-to-final film arrangement. The second had three folding mirrors, which reverted the image, thereby restoring the true head-to-head registration of alternate images. Exposing was simply done by alternately shuttering the lenses as the print film was indexed ahead.

The lenses were Goertz Red-Dot Artars — process lenses ordered directly from the catalog.[33]

SAMOS Program 201 (E6)

In March 1962, a memo written by NRO Director Joseph Charyk and signed by Deputy Secretary of Defense Gilpatric ... directed that all DOD space activities be classified Secret. No individual projects were to be identified, nor was any individual — civilian or military — to be identified with any specific project."[34]

33 Descriptive material provided by Keim, who should be credited with the original concept and solution to the problem.
34 J.T. Richelson, "America's Secret Eyes in Space," p. 65

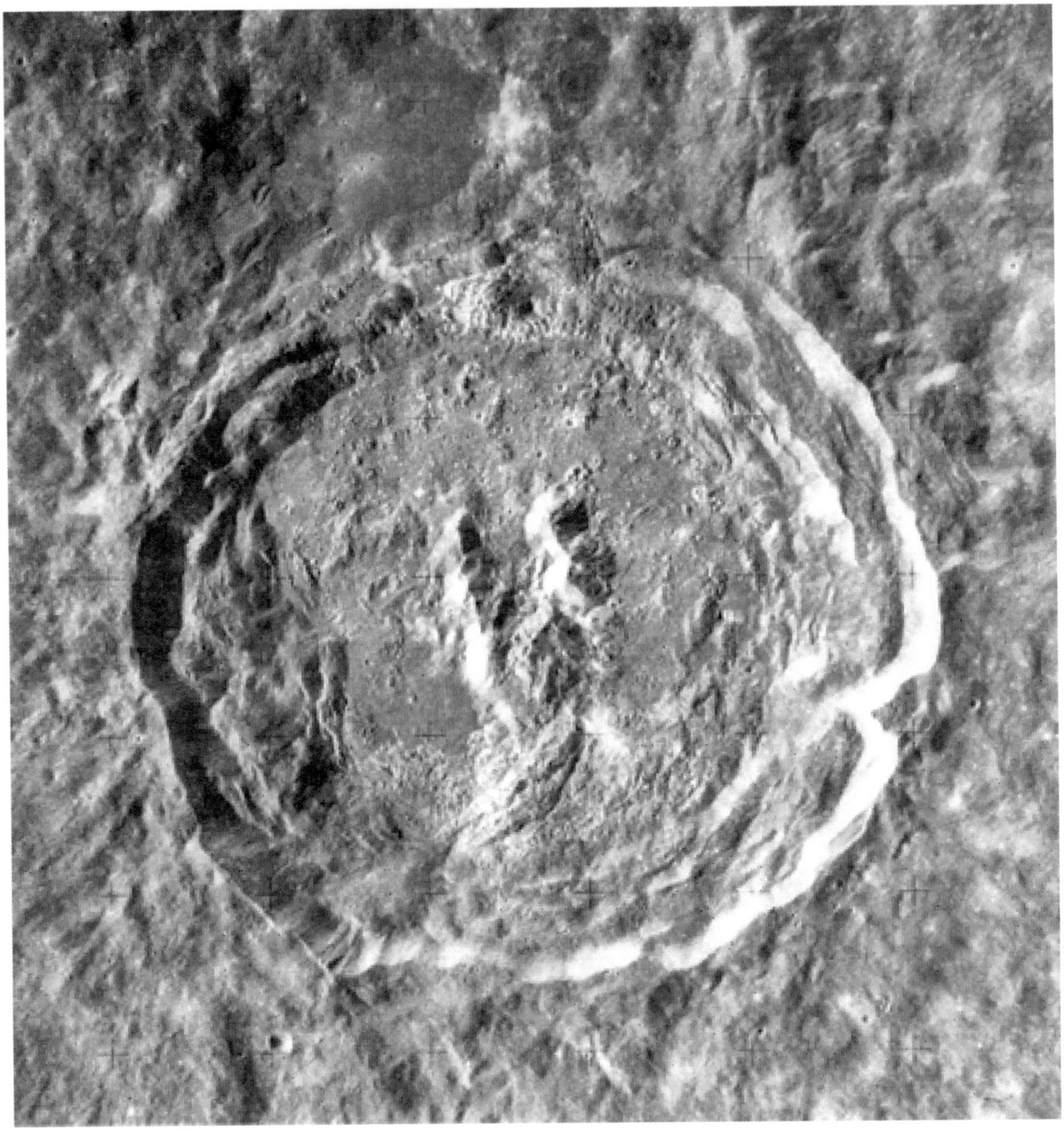

***Photograph 20: Apollo 16 Image of* King Crater**
An Apollo 16 image of the King crater. Shortly after Lunar Orbiter maps of the back of the moon became available, craters were named in honor of people of importance in the history of Astronomy.

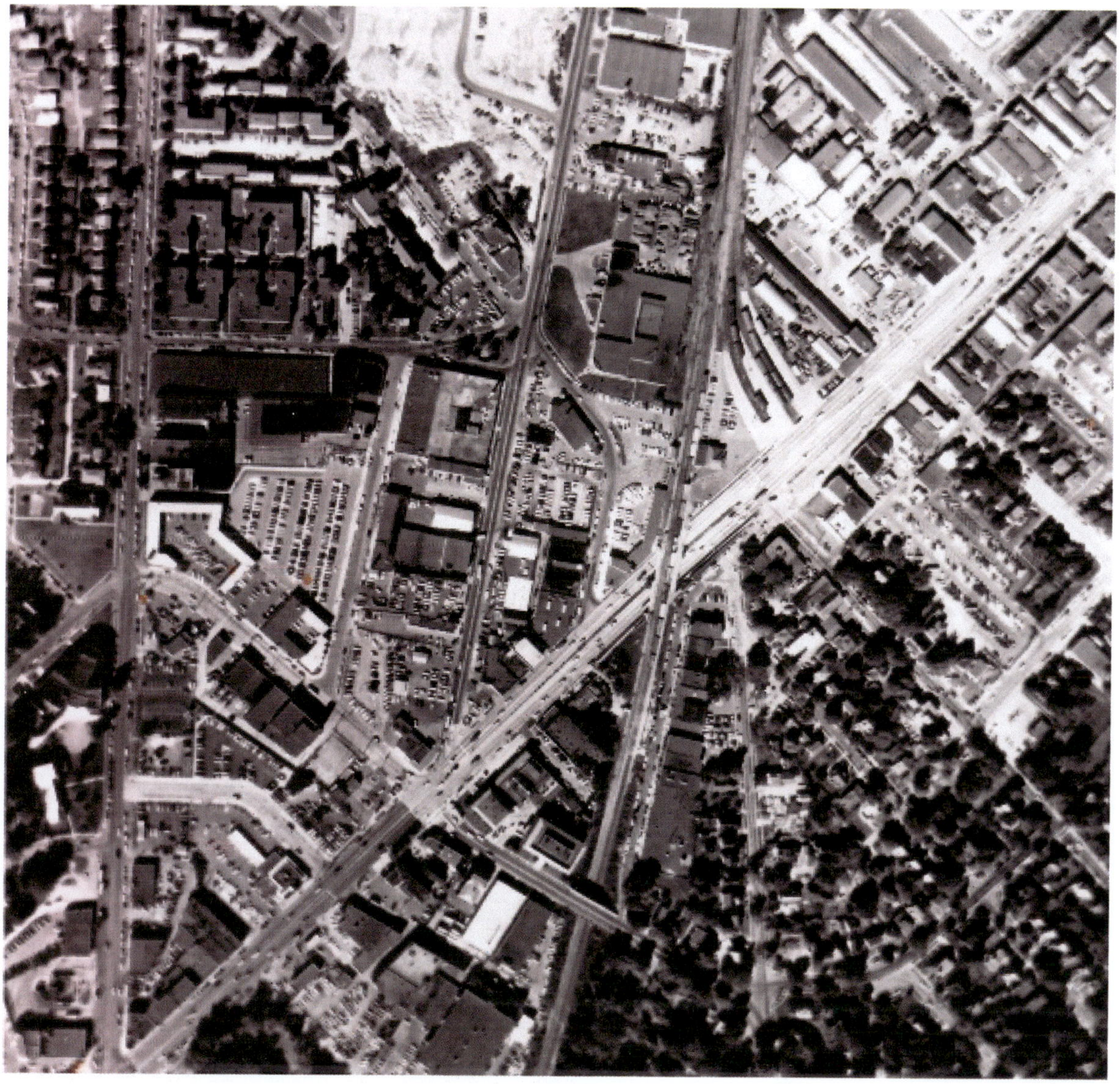

***Photograph 21:* High Altitude Aerial Photography Simulation**
A simulation taken from high altitude aerial photography — ground resolution 2 to 3 feet.

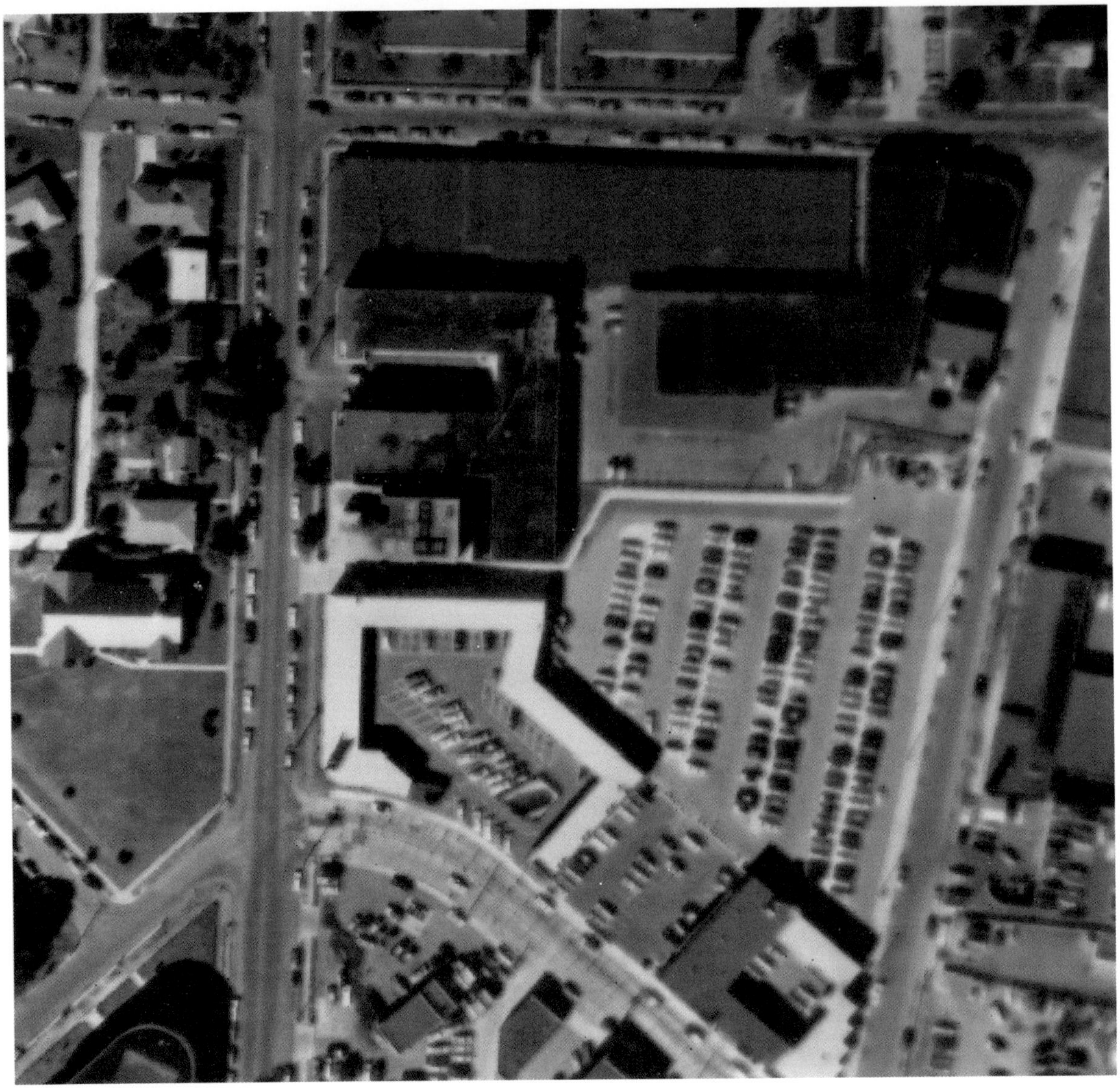

Phtograph 22: **Simulation Enlargement**
The same simulation enlarged. A better simulation might have included a a test target in the scene.

Time period	Designation	Code name or Nickname	Optics	Notes
1959-1972	KH-1 to KH-4	Corona	Resolution: 7.5 m, 2.75 m, 1.8 m Focal length: 0.6 m	First known series of US spy satellites; photographs returned by film canister ejection.
1960-1962	-	Samos	Res: 30 to 1.5 m Foc len: 0.7 to 1.83 m	Most flights used radio to relay images; some film returns; probably cancelled due to poor-quality imagery.
1961-1964	KH-5	Argon	Res: 140 m Foc len: 76 mm	Film return.
1963	KH-6	Lanyard	Res: 1.8 m Foc len: 1.67 m	Shortlived operation for imaging a specific site; used a camera from the Samos program; film return.
1963-1967	KH-7	Gambit	Res: 0.46 m	Film return.
1966-1984	KH-8	Gambit	Res: 0.5 m	Film return.
1971-1986	KH-9	Hexagon Big Bird	Res: 0.30 m	Film return.

Figure 8: **U.S. Reconnaissance Satellite Data**
A Wikipedia web page posted in the 90's or before. An interesting summary of U.S. reconnaissance satellite data, apparently gleaned from various sources. It was the resolution (.5 meters) given for KH-8, lower than both KH-7 and KH-9, that began the author's search for answers.

United States Satellites
- Lacrosse/Onyx
- Misty/Zirconic
- Samos
- Quasar
- Vela
- Vortex/Chalet
- Key Hole (KH) series of Imaging Satellites

Soviet Union
- Cosmos
- Rorsat
- Almaz
- Yantar
- Zenit

Germany
SAR-Lupe 1-5

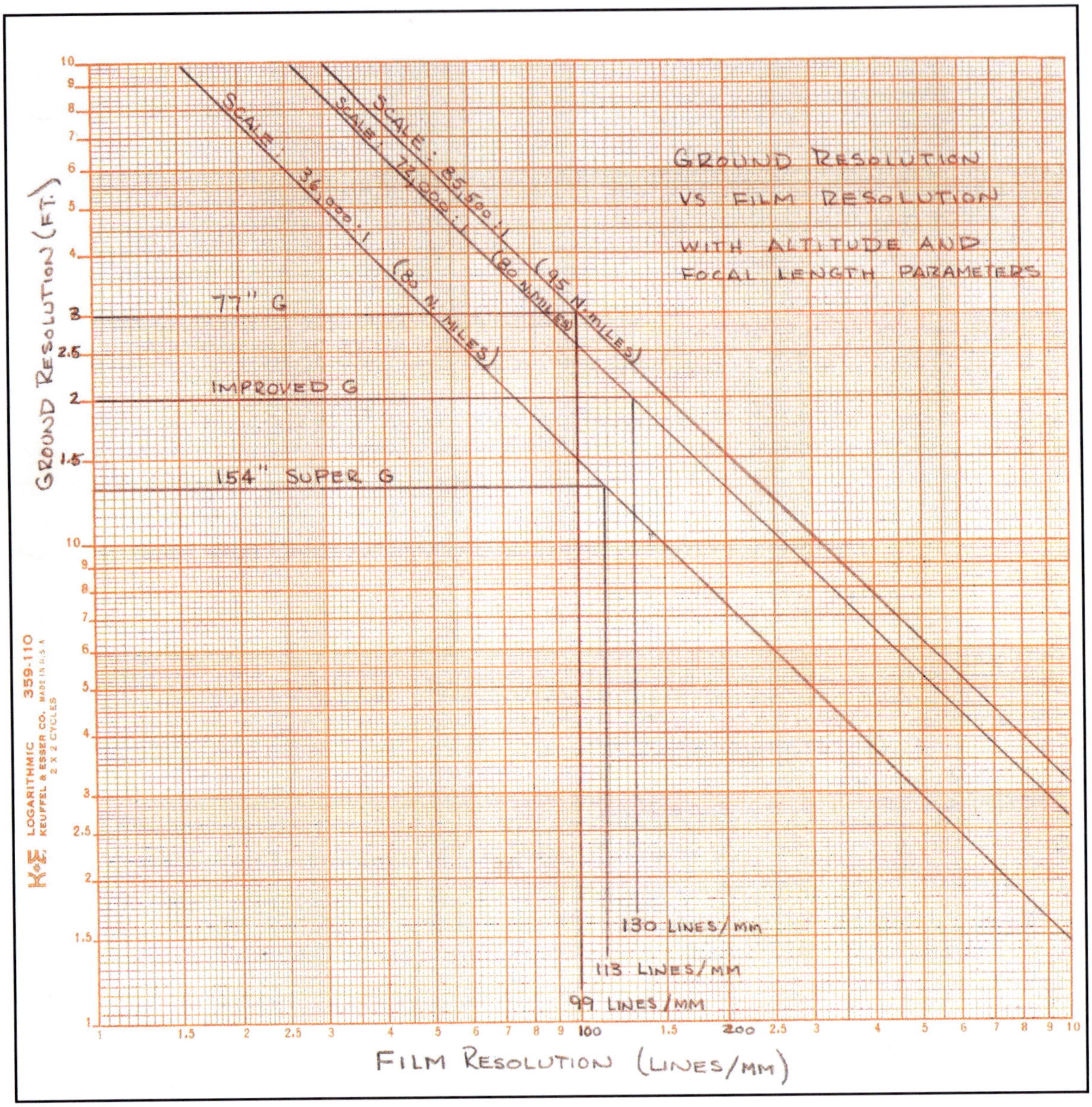

Figure 9: **Historic, 1960's Artifact**

A most interesting, historic artifact, dating back to Kodak's 1960 proposal for a 77 inch lens and 2 to 3 foot ground resolution. Unfortunately, the chart is misleading because the film resolution axis should have been labeled lens-film resolution. Although the resolution or resolving power of the film to be used on GAMBIT was considerably higher, film characteristics were Kodak proprietary.

However, strict security measures were in place, under Air Force technical management, well before that date.[35] I never thought about it then but I was engaged with the Air Force again after a brief interlude (1946-1958).[36]

The operational phase of SAMOS (that is, the E2 readout surveillance system) began late in 1960 and was terminated in 1962 — returning "not a single useful photograph.[37] Considering the overall complexity of the system, its surprising that the camera and read-out system did work.[38] Satellite instability and the attitude control system certainly contributed to poor imagery.

The system had an auxiliary imaging path to look at the horizons. Prisms and windows provided images of the forward and sideward horizons both on film and position sensors. However, horizon imagery in visible light was not well enough defined to allow accurate determination of the system attitude in space.

Only brief mention is made of the first SAMOS film recovery system, Project 101B (E5).

> "Eighteen launches of the SAMOS, KH-4[39] CORONA, KH-5 Argon and KH-6 LANYARD were conducted throughout 1962, using Thor-Agena B or 44 Thor-Agena D rockets. However, each program was plagued with high failure rates.[40]"

Richelson goes on to say that CORONA failed to produce useful imagery throughout 1962. Is it possible that there were a few successful SAMOS missions during that period using the 36-inch R-327 lens? The evidence for this is in the Wikipedia webpage [41] (See Figure 8), which gives ground resolution of 1.5 meters (4.9 feet), which was the best to date.

Again there is little information on whether any of these lens-camera payloads produced useful imagery since all of the SAMOS missions are lumped together. The only evidence with respect to ground resolution is in the Wikipedia web page under "optics." The focal lengths given are questionable and bear little resemblance to the known SAMOS lenses. The KH-6 Lanyard satellite used the "basic E-5 SAMOS camera" (recovery high magnification) but with what lens? Resolution of 1.8 m is given with the focal length 1.67 m (a 66-inch lens?) Did they substitute a 66-inch lens for the proven 36-in SAMOS lens or could it have been an even larger diameter lens? Not likely.

What I deduce from all this is the fact that the lens used for the Lanyard satellite, in all probability, was Kingslake's 36-inch lens, which produced ground resolution readings of 1.5 to 1.8 meters both on SAMOS and the Lanyard programs!

[35] Note: Under the Air Force's Western Development Division, the WS-117L program existed at least as early as February 1958.

[36] In WW II I served in the Air Force from January '43 to February '46.

[37] J.T.Richelson, p. 68

[38] Ibid. p.68

[39] Note: the KEYHOLE (KH) designations came into use in 1962.

[40] J.T.Richelson, p.66

[41] Wikipedia.org/spy satellites

8

KH-7 GAMBIT

In 1961 the highly classified SAMOS and GAMBIT programs were located within Research and Engineering at Lincoln Plant and Hawk-Eye. As director of Kodak's Apparatus Research and Development Organization, including the Research and Engineering Division, A.B. Simmons was responsible for allocating space for the GAMBIT program. As work began, areas were being prepared at Hawk Eye for optics manufacturing in the basement of the new building and engineering on the third floor of the old building.[42] All GAMBIT program work would be consolidated at Hawk-Eye under Frank Hicks.

Dave Schaeffer, who was head of security, recalled the rather sudden disruption and confusion surrounding the move. Everyone was told to turn in all classified documents and pack everything they had been working on in boxes. It looked like the program had been canceled but the next day we were to show up at Hawk-Eye where everything was returned.

The GAMBIT program was essentially the development of a scaled-up version of the R-327 formula lens. Not having seen the proposal, I believe it included not only the projected ground resolution but also simulations taken from high altitude photography (Photograph 21 and 22). Optics of this size were well beyond the state-of-the-art in the early 60's.[43] They represented a significant challenge together with that of the camera. The focal length more than doubled, from 36 inch to 77 inch, and the aperture increased from 9 inch to 20 inch. The actual size of the primary mirror was 28 inch and the stereo mirror 33 inches. Film width went from 70mm to 5 inch.

If defocus and smear effects are neglected, ground resolution estimates or projections can easily be made based on actual film plane measurements of resolution in lines per millimeter. I had forgotten having made such projections until I came across the original chart- a most interesting, historic artifact (Figure 9). The in-factory measurements of resolution are indicated together with the projected ground resolution in feet with altitude and focal length as parameters. The letter G was an internal designation, which began with the SAMOS program (E6) and later given the GAMBIT designation. The improved G represented the introduction of the new Estar based film.

Note that the ground resolution projections of 3 and 2 feet indicated for the "77-inch G" and the "improved G," are in good agreement with the published values in the website maintained by John Pike, GlobalSecurity.org: "Best resolution was initially four ft. (1.2 meters) on ground; by 1966, best resolution improved to two ft. (.6 meters)." Resolution of 1.5 foot (.46 m) is the number given in the Wikipedia website,[44] (Figure 8), which corresponds to film-plane resolution of 200 lines/mm, and the bird flying at 95 nautical miles.

[42]As I reflect back, we were so isolated from other activities in the building, I was totally unaware of my old Department 35 Engineering on the fourth floor directly above us. Bridgehead was on the fifth floor.

[43]The only tools available to the optician were an 8 inch diameter test plate and a Foucault Knife Edge tester. Lens bench tests, using an Autocollimating Point Source Microscope, provided a check for astigmatism.

[44]http://en.wikipedia .org, Spy satellite

So why were the film resolution numbers, 99 and 130 lines/mm, so low? Unfortunately, the chart is misleading. As indicated above, I believe these numbers represent actual or typical readings from tribar tests on the collimator,[45] which means they were lens-film measurements, not film resolving power.

The resolving power of the film used in those tests, probably one of the Fine Grain Aerial films, was 300 lines/mm; the film speed (AFI) 6-10. The film for the "improved G" was therefore a High Definition Aerial Film with resolving power, 500 lines/mm, and film speed (AFI) 1.6. [46]

In scaling up aperture and focal length (both the 36inch and 77inch lenses were f/4's), it turns out that, regardless of scale, resolution at the film-plane should be the same, if they were manufactured to the same quality. The lens-film resolution on-axis would then have been 275 lines/mm and ground resolution 1 foot.

As the design phase of the Gambit-1 program (referred to internally as the R program) wound down, optical design and configuration studies were initiated (the Valley program) on a possible follow-on. The projected ground resolution for a 2x scale-up of the G lens is shown for the 154 inch Super G (see the Figure 9). The Lens-film resolution of 113 lines/mm was an average value.

In the expectation that the KH-7 and KH-8 programs as well as Samos may be declassified, I feel justified (for the public record) in disclosing information that may otherwise be lost. In light of the "hundreds of millions and billions of dollars ... spent on (SAMOS) per Richardson,[47]" the Kodak story needs to find its niche in the history of the nation's satellite reconnaissance programs.

[45]There were two 300 inch focal length collimators used for acceptance testing.
[46]See listing of Kodak Aeial Films, Appendix A, Bridgehead, The Story of Its Role
[47]J.T.Richelson, "America's Secret Eyes in Space," p. 68

9

KH-8
The Next Iteration

The success of the NRO directed Corona and Samos film-recovery programs only whetted an insatiable appetite for higher resolution imagery.

"...the Air Force's more ambitious Samos Atlas-Agena reconnaissance project...[48]"

...leaned heavily on Kodak's research and engineering division for the optics and camera systems. The KH-7 GAMBIT program had improved ground resolution from the reported 5 feet on a few Samos missions to 4 feet initially and eventually 2 feet. By August 1963 the proposal to double once again the size of the optics, the primary mirror diameter from 20 inch to 43 inch, was indeed ambitious, calling for a massive build-up in facilities and personnel.

Optics may be a totally foreign, arcane subject for the vast majority of casual readers but for students of optics and for those who have found a livelihood in this somewhat exacting field, the fascination with these near perfect lenses in space is in the details: the design and implementation.

Months of effort went into lens design studies, followed by, or in parallel with, the preparation of a requirements document (and the System Technical Manual or Handbook), then the optics manufacturing and assembly phases — all on a crash program (1964 to first launch July 29, 1966). I'll never forget the day Cal Rogerson, representing Reliability, and I made our presentation to the customer. You didn't admit to being nervous, but these were high-ranking officials. I can't remember if there were introductions — but of course I didn't have the "need to know" who they were. Identities were classified.

The half scale layout of the optics stretched 7 feet across the conference room wall! By that time the KH-7 lenses were in production and the next step in this madness of ever increasing size seemed reasonable — perhaps because it was inevitable. I know I tried to exude confidence but inwardly I had serious reservations about the 43 inch asphere, the massive 58 by 46-inch stereo mirror and whether the Air Force could lift this thing into orbit. It was obvious that the program would call for the development of lightweight mirror blanks. Negotiations with Corning had not yet begun.

Requirements had to be established for the vehicle attitude control system — pointing accuracy and rate of change in Yaw, Pitch and Roll based in part on the analysis of image smear and its effect on image quality.

[48] The Air Force, The Air Force Historical Foundation Publication, p. 230
Published by Hugh Lauter Levin Associiation, Inc. 2002

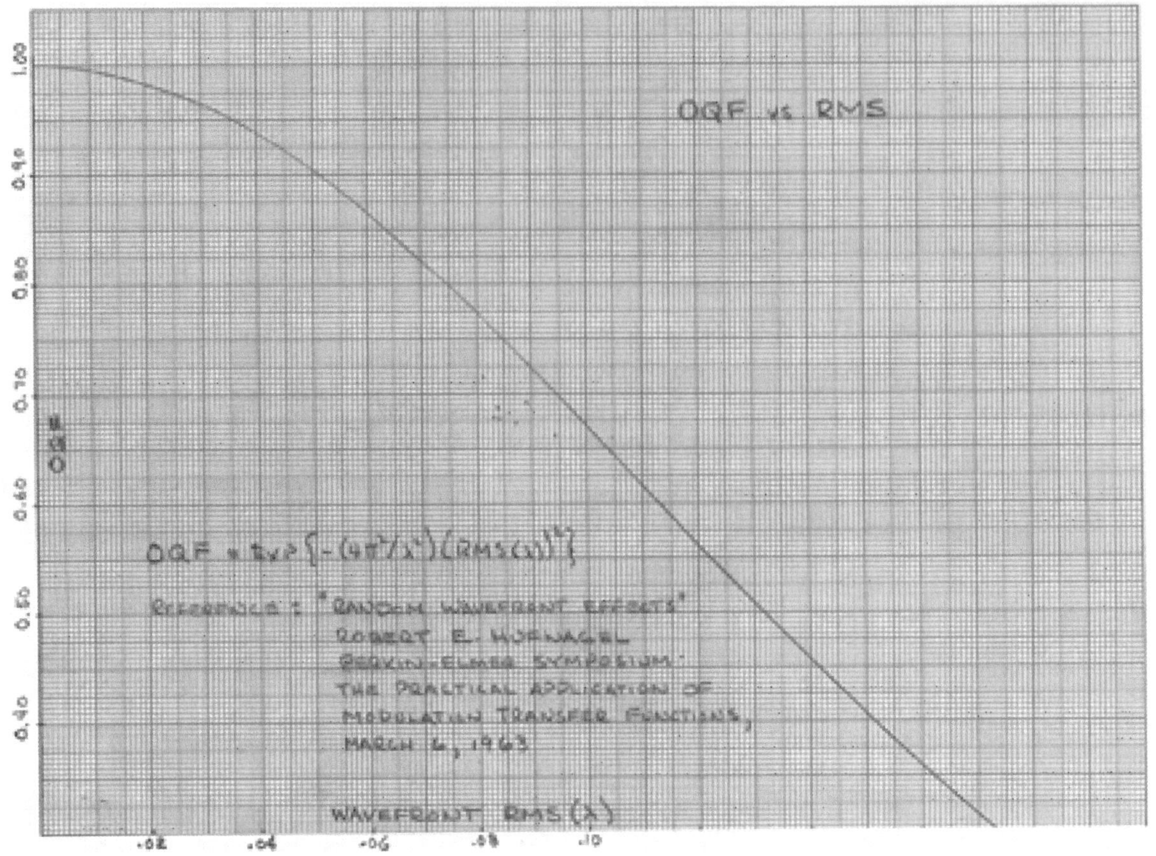

Figure 10: **Hufnagel Formula Plot**
A plot of the Hufnagel formula which defines OQF as a function of the RMS wave front error. The March 6, 1963 date of this Perkin Elmer Symposium coincides roughly with the use of modulation transfer functions (MTF) at Kodak in system analysis.

Image Quality Analysis

As it is the lens designer's goal in life to minimize lens aberrations, the goal of the optical engineer, what he brings to the table, is to evaluate and minimize the effects of the lens manufacturing process. Back in the early 50's Charlie Bridge[49] and I conducted a study on the T-46 rangefinder program to demonstrate experimentally the effect of aperture and axial astigmatism on the theoretical limit of resolution. 1951 Air Force Tri-bar Targets were used.

It was our job to establish surface tolerances, measured in wavelengths of light or fractions there of, and the system requirement. Each surface contributes to the overall optical quality just as it does for a lens.[50] What the study demonstrated was the fact that for visual optical systems there is no practical value in requiring the astigmatism to be better than 1/4 wave peak-to-valley.

Similarly for very large aperture lenses and their component mirror surfaces the criteria had to be stated in terms of the RMS deviation from the perfect surface or wave front. There was no value in reducing wave front errors below about .10 RMS. The lens was, for all practical purposes, diffraction limited. For the Gambit-3 lens and stereo mirror assembly the single-pass wave front tolerance was set at .095 RMS or 70 percent OQF.

The relationship between the optical quality factor (or OQF) and the measured RMS wave front of a lens is given by a formula attributed to Robert E. Hufnagel (Figure 10). This was the math model for establishing the lens specification.

Without undo elaboration, mathematicians on the program developed the Sine-wave Response or Modulation Transfer Functions (MTF), as a technique for evaluating the effect of system variables that contribute to a loss of image quality. For example, the sign wave response curves for a diffraction-limited lens in and out-of-focus with a central rectangular obstruction closely model the KH-8 lens response to a defect of focus (Figure 11). The response function for the lens could then be combined with that of the film to predict modulation and lens-film resolution at the image plane. However, I have no evidence that this was actually done.

Lens Design Activity

From the very beginning of these reconnaissance programs, lens design studies were in progress both at Kodak, the Institute of Optics and Tropel, an offshoot of the Institute. A few of these designs, or formula sheets, dating back to the 50's and early 60's, have survived, stuck away in my files, and deserve special mention. As far as I know, the first two designs were Kingslake's: the 36-inch R-327 designed in December 1948 and the 120-inch R-360 designed a year later, December 1959. The R-327 formula, as mentioned earlier, was the design for Kodak's Samos programs first in a 36-inch focal length (E-5 and E-6) and then redesigned for the scaled-up 77-inch version (KH-7). But what about the 120-inch R-360 formula, Kingslake's other lens formula? Was it ever implemented? It had long been a mystery until just recently. It didn't seem to fit into either the Corona or the Samos programs. This is what I think happened. For me, at

[49]Charles Bridge, a Dept. 35 engineer, who later left the company for Litton Industries.

[50]Note: The surface contributions of refractive elements in these large lenses were quite small.

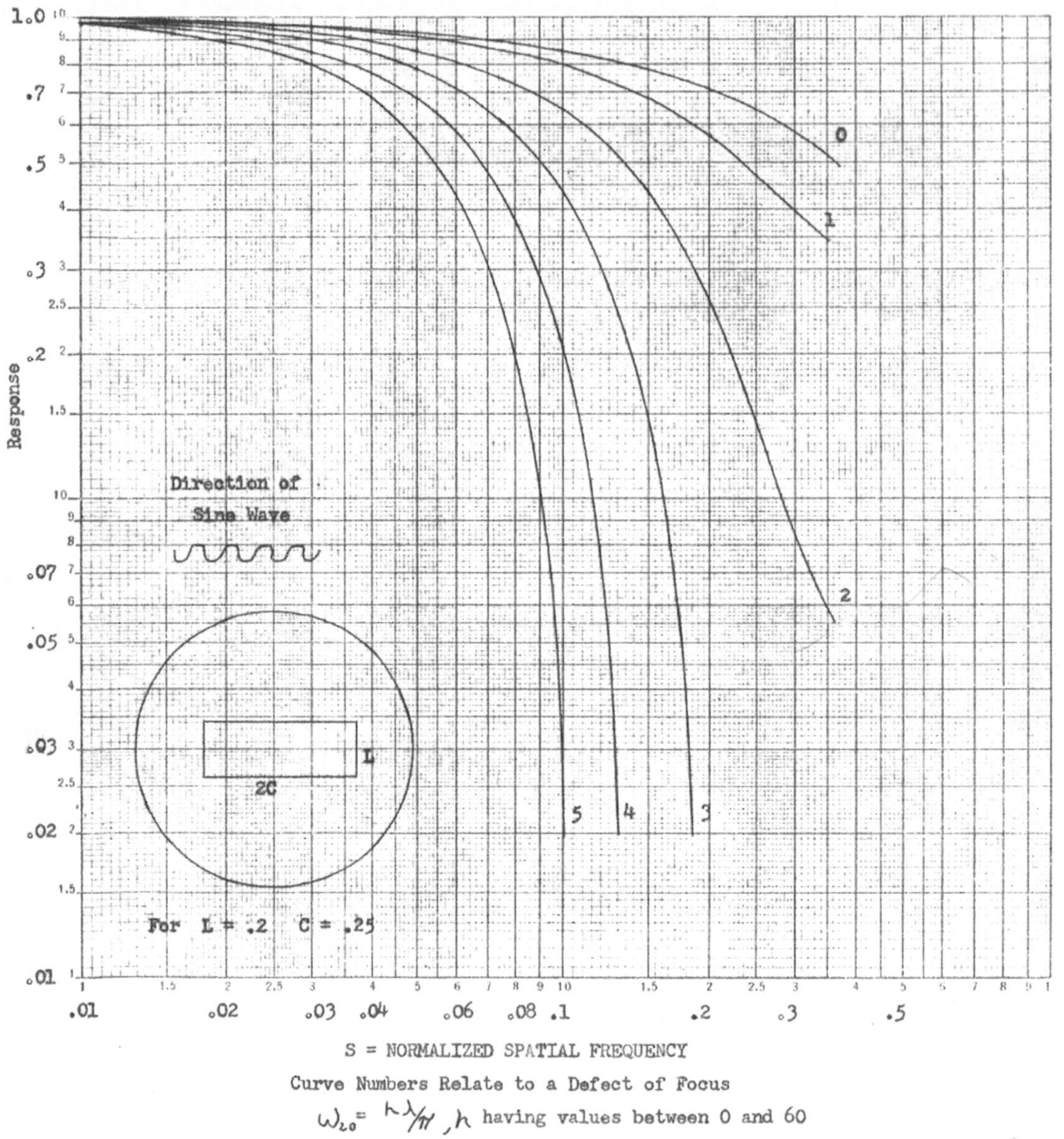

Figure 11: **Diffraction Limited Lens Response Graph**
Graph of the out-of-focus sine wave response of a centrally obstructed diffraction limited lens. This was one of about 16 graphs compiled by the R & E Systems Group.

least, it explains the mystery of the 120-inch lens as well as the fact that these are the only two existing Kingslake formulas.

I was first aware of the Institute's involvement in a classified program reading through the magnificent volume, "*A Jewel in the Crown*," 75th Anniversary Essays, Institute of Optics, University of Rochester. There on page 97 was this interesting reference:

> "Of great interest was the development of the 36-inch reflecting telescope for an orbiting satellite."

And so I suspect these two Kingslake formulas, the 36-inch and the 120-inch, were designed under contract to either the Rand Corporation or Itek — during the inception of Corona.

Here is the corroborating evidence, borrowing from Richelson's "gripping" account of that time in U.S. history, October 3, 1957, Sputnik II, and the sequence of formal recommendations from Rand Corp. "Fritz" Oder represented the Air Force's 117L program office.

> "...November 12, RAND formally recommended to the Air Staff development of a recoverable reconnaissance satellite. RAND's president, Frank Collbohm, provided a cover letter for [their] recommendation, ***An Earlier Reconnaissance Satellite System.*** He wrote that:
>
> "In the light of recent events, RAND has reviewed national and military intelligence problems, existing and proposed reconnaissance systems, and in particular, the current USAF satellite reconnaissance program (WS-117L)... The systems proposed in this recommendation differ substantially from the 117L system concept.
>
> The proposed systems feature a spin-stabilized payload stage. They use a transverse panoramic camera of essentially conventional design.
>
> The first of the proposed systems uses a 12-inch camera carrying 500 feet of 5-inch wide film ... it will provide sharp photographs of about 60-ft. ground resolution.
>
> The next, more sophisticated, system would use a 36-inch camera, carry much more film, and do more detailed reconnaissance with a ground resolution of about 20 feet.
>
> A third system -- undoubtedly requiring Atlas-type boosting -- would use a 120-inch camera and would have very large film capacity. This system will be able to accomplish very high quality photo reconnaissance and, most important, would do it better than any Air Force system now in development or will be able to do in the 1960's.[51]"

So it was this Rand recommendation that provided the overall plan for a parallel satellite reconnaissance program at Itek, a newly formed company in the Boston area. And here was the reference to a "120-inch camera" or, as it was understood to be, a 120-inch focal length lens-the answer to the mystery of Kingslake's 120-inch lens formula.

[51] J.T.Richelson, "America's Secret Eyes in Space", p 21

It's interesting to note that Kingslake's contribution to these programs was in the form of two lens formulas, the 36-inch design used on the Kodak's E-5 and E-6 Samos programs and the 120-inch design that went to Itek. To my knowledge that 120-inch formula was never converted into hardware because it would have duplicated efforts at Kodak.

To complete the story, as I know it, of lens designs in connection with Corona, Itek went from the suggested 12-inch to 24-inch f/5 Tessars for their transverse-oriented panoramic cameras. The other program that Itek may have been involved with was the KH-6 payload Lanyard, which "used a camera from Samos" with an unidentified 1.67m (66-inch) focal length lens. There were five of these mysterious systems built with Samos cameras.[52]

Design Activity for the KH-8 (Gambit) Program

Perhaps the most significant, even crucial, contribution made by the Institute of Optics was in the area of lens design, in the training of lens designers and in establishing the small company in Fairport called Tropel. Interesting enough it came into existence out of work performed for the Department of Defense.

> "During World War II, the Institute worked for the Department of Defense on image-stabilized binoculars. Some years later, interest in this project grew, but the Institute no longer wanted to participate. In 1953, Robert Hopkins proposed to Jim Anderson and Jack Evans that the three of them start a company to do this work. Thus Tropel was born... The first full-time employee was John Buzawa,[53] also of the Institute of Optics.[54]"

My first contact with Buzawa, Director of Optical Design, was some time in 1962 when Kodak initiated a contract with Tropel for lens design studies. His final report titled, Optical Design Studies of Catadioptric Systems, dated July 5, 1963, summarized the performance specifics (spot diagram, energy distribution, and geometrical frequency response) of a Maksutov design similar to Kingslake's 36-inch R-327 formula. As he indicates in the report, "In order to obtain the actual frequency response it would be necessary to combine this curve [the geometrical frequency response] with the proper pupil function to take into account the effects of diffraction since the geometrical errors are quite small." In other words it was a diffraction-limited lens in the normalized focal length of 100 units. Unfortunately, the feasibility of building such a lens to the scale required made it impractical for the Gambit-3 program.

Keim's Report

The debate that essentially began with the early configuration studies, is best characterized by the most dire predictions coming from R.E.Keim, based on Gambit experience. Just to offer a little perspective to the daunting task that lay ahead, the following excerpts from Keim's November 1963 report, "Areas of Critical Difficulty in the 160-inch f/3.7 Lens[55]" set the tone:

[52]J.T.Richelson, America's Secret Eye's in Space, p.68

[53]M. John Buzawa, B.S., 1952; M.S., 1954

[54]A Jewel in the Crown, 75th Anniversary Essays, The Institute of Optics, University of Rochester

[55]During the early study phase for the KH-8 program.

"***Configuration:*** Lens (without flat mirror) can not be easily tested as an independent unit Camera location on opposite side of flat mirror from primary requires elaborate mechanical linkage or frame

Formula: All spherical aberration correction is achieved by aspherizing the primary mirror. This is a very critical situation, considering formula tolerances

Manufacturability: Primary mirror
The aspheric surface demands 4 times the surface accuracy as an aspheric refracting element would require Because of the location of the stop at the primary its surface must be within one tenth wave from center to edge.
The Flat Mirror
The asymmetric hole[56] will cause asymmetric polishing over the entire surface. Whether the defect can be kept within tolerance is questionable.

Performance Level: The manufacturing losses can be expressed as a % loss of the Modulation Transfer Function at 150-200 lines/mm ...It is unrealistic to claim a 30% maximum loss... when one considers that its beam area is about 4 times that of the 77 inch [E-6] and 25 times that of the E-2 [E-5] lens."

Keim's optics configuration utilized a scale-up of Kingslake's R-360 120-inch formula- the one I puzzled over and concluded it had its origin in the Rand Corp. recommendations. It did not require a hole in the stereo mirror but the formula had a limited field and "in its present state very poor correction".

Another approach, also initiated at Tropel, inspired by Kingslake's 120 inch lens formula, was to make the primary mirror aspheric eliminating the large corrector elements in front. Four field flattener elements were placed close to the image plane.

This design study resulted in the R-361 formula, dated January 1964.[57] This was the lens formula[58] used on at least twenty of the early Gambit 3 payloads prior to a major redesign. Layouts showed the field flattener and platen brought out to the side of the large 58 by 46-inch elliptical stereo mirror. This necessitated a lengthy film path with ninety-degree turns. In addition the back of the stereo mirror had to be beveled to fit within the optics barrel.

It was two years after the first KH-8 launch that these serious problems came to a head not the least of which was the performance level only slightly better than that of KH-7! By late 1967 the inadequacy of the Gambit-3 camera system remained an unsolved problem. It did not obtain the resolution originally specified."

[56] Its interesting to note that the hole in the stereo mirror was considered but rejected.
[57] Keim makes the statement that Kingslake's R-360 formula was poorly corrected off-axis, which raises the question how well was the R-361 formula corrected off-axis? I never had access to the design data!
[58] The focal length scaled-up to 160 inch

EASTMAN KODAK COMPANY - - HAWK-EYE WORKS

Formula: R-360 — OPTICAL DESIGN DEPARTMENT — Aperture: f/4

E.F.$_D$ = 120.13 — 120-inch Catadioptric System — B.F.$_D$ = 1.202

Ang. Field: ±1° (image diag. = 4.2") — (Dimensions in inches, aberrations in thousandths.)

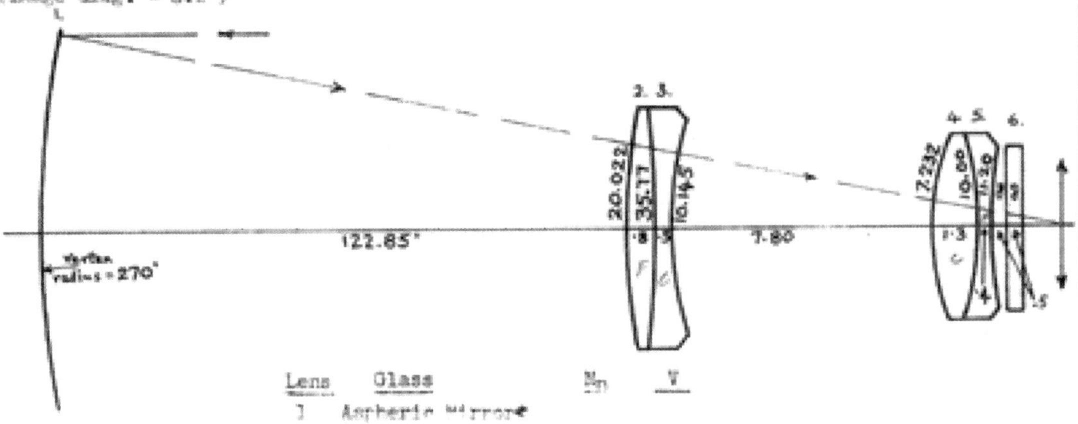

Lens	Glass	N_D	V
1	Aspheric Mirror*		
2	EK-210	1.734	51.0
3	C-1	1.523	58.6
4	DBC-1	1.611	58.8
5	DF-2	1.617	36.6
6	window or filter	1.520	58.2

Surface	Cl.Aper.	Axial Beam
1	30.00	30.00
2	6.98	2.71
3	6.90	2.57
4	6.51	2.44
5	5.40	.80
6	5.23	.58
7	4.85	.51
8	4.76	.39
9	4.65	.30

Aper.	C	D	F	O.S.C.
f/4	-.04	-.03	-.42	.00016
f/5.7	-.17	-.04	-.12	.00008
Parax.	-.25	00	+.20	

Field	Dist.	X'_s	X'_t	Lateral Color F-D	Lateral Color C-D
1°	19.3	-.23	-1.31	+.07	+.12
0°40'	5.8	+.18	+ .80	-.17	+.16

*x = ay² + by⁴ + cy⁶

a = .001 851 852
b = -.000 000 00 142
c = -.000 000 000 0001

Rim Rays

A	ΔH' at 1°	ΔH' at 0°40'
15	-.82	+.20
10.5	-.29	+.20
0	00	00
-10.5	+.16	+.15
-15	+.16	+.28

Petzval Sum (100) = .00036

DESIGNER R.K.	ASSIGN. 1226
COMPUTER R.K.	F-83 File 227
DRAWING BY R.K.	DATE Dec 3, 59

HE 690

***Formula 4:* Kingslake's Mystery Formula**

Kingslake's mystery formula, a 120-inch Catadioptric System for Lincoln Plant, with its origin in a Rand Corporation report. Designed in 1959, it was later scaled up to a 160-inch lens for the long focal length Gambit-3 proposal.

I had known this back in 1967 when reports filtered down. Then in the late nineties the Wikipedia web page listed the Gambit-3 resolution at .5 meters or 1.5 feet as slightly worse than Gambit. I knew of course that for the early missions it had to be the lens formula and suspected that there was a sharp fall off in image quality. There were also these major concerns in supporting and polishing the stereo mirror in addition to feeding film to the platen and back to the camera.

The proposed alternative optics configuration was literally a break-through — borrowed from Lewis Carrol's "Alice in Wonderland".

Through the Looking-Glass

The large elliptical stereo mirror would have a large rectangular hole in the center of it allowing the convergent beam to pass straight through with image plane and camera directly to the back of it. This greatly simplified film handling. At the same time it was decided to drop the center-line of the optical system. This eliminated the need for the bevels and the stereo mirrors were successfully ground and polished using a removable plug in the hole. Polishing the aspheric primary mirrors required the use of a new Null corrector.[59]

More important was the possibility of redesigning the lens — an advantage that was not fully realized until the new R-5 formula was developed. The "through the looking glass "approach gave the lens designer more or less complete freedom in the placement of lens elements. The final optical design evaluation report from Tropel on the R-5 lens design was dated September 10, 1968, which meant the complete redesign and implementation of the lens and film handling system occurred during the months that followed, probably to the end of 1969. That year there were eight KH-8 launches with the introduction of the new R-5 lens configuration in January of 1970.

To recap the situation at the start of the program, the only configuration that made sense was one based on the R-361 formula, dated 1-27-64, designed by Tropel. Apparently it was stipulated in the design study contract that all refractive elements had to be within a certain distance from the image plane. The formula issued by the Optical Design Department was for a 100-unit focal length lens. It was never evaluated at the actual 160-inch focal length because Tropel did not have clearance. Presumably the design passed through the Optical Design Department and had Kingslake's approval. In retrospect it was not at all certain that the lens formula provided adequate quality off-axis. (See Appendix B)

Kodak had no experience polishing large aspheric mirrors and there were doubts about fabricating and then polishing the even larger stereo mirror with the large 45 degree bevels.[60] Negotiations with Corning for light-weighting the mirror blanks had not yet begun and the proposed low thermal coefficient ceramic called Pyroceram would not have been thermally stable in space. In other words, the program was "in deep yogurt" to quote Stu Lambers who had overall responsibility for integrating the camera/optics subsystem.

[59] The Null Corrector consisted of two large lens elements placed in the convergent beam during interferometer testing.
[60] The development of interferometer testing was absolutely critical to the success of the program.

EASTMAN KODAK COMPANY- A. & O. DIVISION - OPTICAL DESIGN DEPARTMENT

FORMULA: R-361

E.F.$_d$ = 100.00

Ang. Field ± 1.525°

CATADIOPTRIC LENS

450 - 700 mμ

Aper: F/3.68

B.F.$_d$ = 12.566

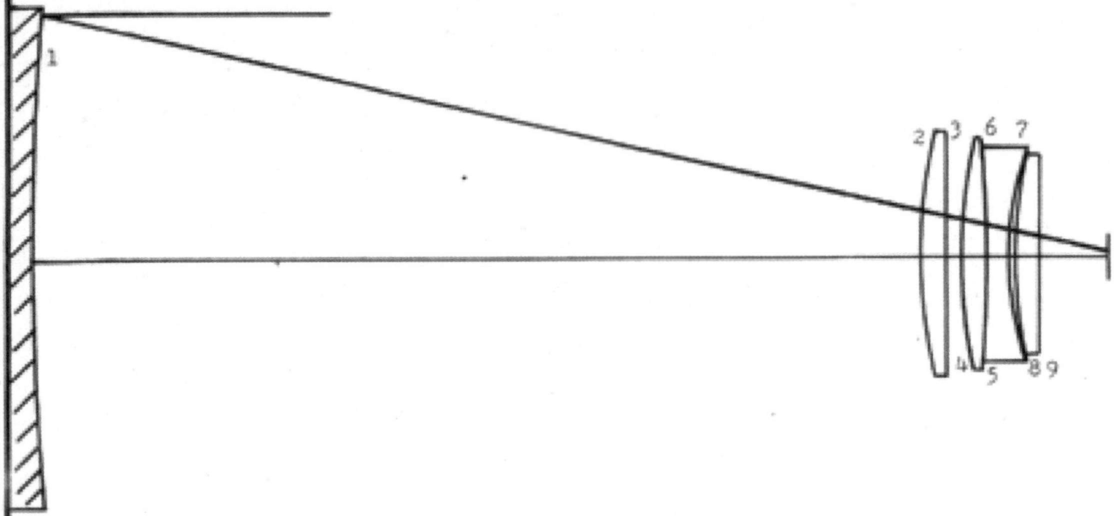

Surf.	Cl.Aper.	Axial Beam	Radius	Thickness	Glass	N_D	V
1(aspheric) (mirror)	27.18	27.18	239.7121*				
				99.832		1.0	
2	9.40	4.58	26.3901				
				.900	K-7	1.51112	60.49
3	9.30	4.42	301.5790				
				.822		1.0	
4	8.90	4.12	21.1851				
				.900	K-7	1.51112	60.49
5	8.78	3.92	-107.9125				
				.010			
6	8.74	3.90	-171.7143				
				.800	LaK-10	1.720	50.31
7	8.26	3.68	12.3478				
				.473		1.0	
8	8.24	3.60	26.2926				
				.900	K-7	1.51112	60.49
9	8.18	3.46	-228.3308				

$$*\text{Sag} = \frac{cy^2}{1+\sqrt{1-c^2y^2}} - 1.8381\times 10^{-8}y^4 - 7.61\times 10^{-13}y^6$$

Where y = height and $c = \frac{1}{r}$

DESIGNER Tropel	ASSIGN.
COMPUTER	
DRAWING BY JV	DATE 1-27-64

AO 778(REV.9-61)

Formula 5: **100-Inch Tropel Formula**

First of two lens designs used on Gambit-3. This 100-inch Tropel formula was later scaled to the 160-inch focal length of the actual Gambit-3 lens.

The original proposal to the customer did not include the hole in the stereo mirror since that decision was made almost three years after the start of the program. But those were the challenges and innovative changes that had to be made throughout the program to guarantee success.

The Key to Success

Of all the challenges faced at the start of the program, the problem presented by the lens design was perhaps the most intractable. The formula was literally the key to the success of the entire program. For a better perspective consider the question: How good did the lens formula need to be? The answer was either simple, or quite complicated depending on who was vitally concerned with the answer. For the customer it was the expectation that resolution of objects on the ground, (in feet or inches) would exceed that of Gambit 1 (KH-7) by a factor of two.

Lens formulas were simply a means to that end. It was not clear at first whether either of the available formulas would fully meet the requirement. If one thinks of the reconnaissance satellite as "America's Secret Eyes in Space," then the job of the optical engineer was not unlike that of the optician who fits the customer with glasses. He needs to know the prescription, the lens formula, and whether it is consistent with providing 20/20 vision. As visual acuity of the eye is limited by diffraction to about one minute of arc, the acuity of "America's Eye in Space," the Gambit 3 lens, was limited by diffraction to .11 arc seconds. That meant that the geometrical errors in the formula should be less than .05 arc seconds or one half the size of the Airy Disc.[61]

I should repeat that at that time it was standard practice for the Optical Design Department to issue a formula and then have a "formula sample" lens built to verify the design before the lens is put in production. I was given the task of evaluating the 160-inch breadboard lens but had no access to the design study report submitted by Tropel. Although I was not permitted to keep test results, I know there were on-axis double-pass interferometric tests, however, off-axis resolution measurements must have been a real challenge to explain.

Computers revolutionized lens design, first with programs for tracing rays and ultimately for the automatic correction of aberrations- automatic lens design programs. Throughout the 50's, ray trace programs provided the lens designer a powerful tool for depicting the geometrical image, the spot diagram, which in turn could be converted to a geometrical sine-wave-response curve. However, the advent of automatic lens design programs in the 60's, brought to bear the full power of the computer in lens design. These were the most sophisticated computer programs known to man, at least at that time. They were indispensible to the evolution of lens design and "America's Eyes in Space."

My notes from a 1962 summer school course at the Institute of Optics included two reports dealing with automatic design procedures: one by G.H.Spencer entitled "A General Linearization Method for Automatic Lens correction" and the other by Dr. R.E. Hopkins which gave sample calculations using the ORDEALS program. These were little more than early

[61]The central spot of light in the diffraction pattern of a point image- so familiar to students of optics and amateur astronomers looking for the holy grail at the end of their polishing.

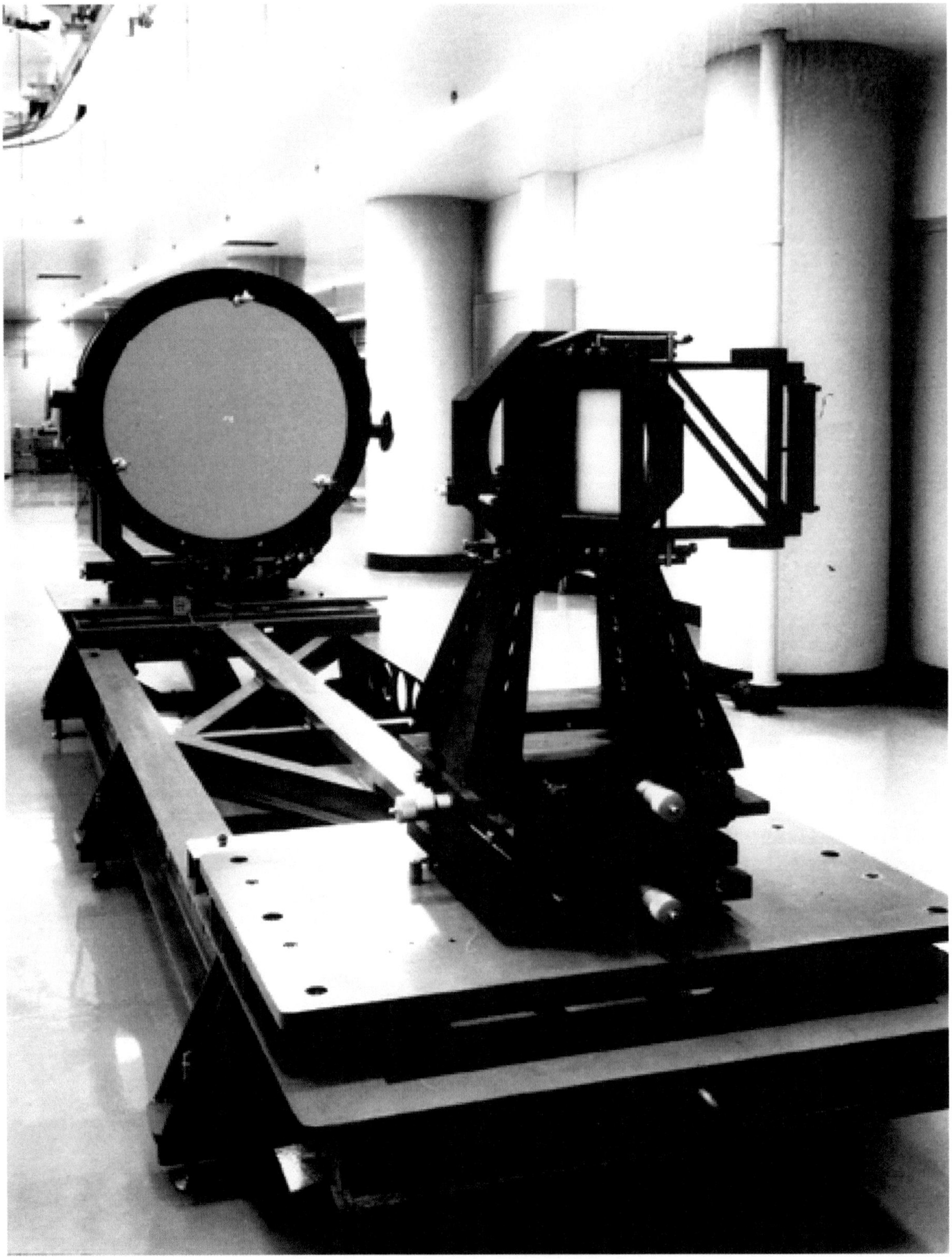

Photograph 23: **R-361 Lens Breadboard**
Breadboard of the R-361 lens formula. The design, procurement and set up of this massive fixture became a high priority at the start of the program.

demonstrations of programs written for the just acquired IBM 7070. Its doubtful that either of Kingslake's 36-inch or 120-inch lens formulas had the benefit of automatic design.

Tropel Enters the Picture

With the expanding need for lens design work, contracts were initiated with Tropel. The first of these was a design study by John Buzawa titled, "Optical Design of Catadioptric Systems." A final report, dated July 5, 1963, provided design data for a Maksutov type lens with a focal length of 100 units, f/3.5 relative aperture and a semi-field angle of 4.2 degrees. The primary mirror was aspheric and the layout showed a split front meniscus, presumably to minimize weight in handling.

From the standpoint of the optical systems engineer this report marked a significant advance in optical system analysis with the ability to combine the geometric response function of the lens with other contributors that degrade image quality.

In my attempt to reconstruct the sequence of events, this study represents something of a mystery since the Samos programs, (E-2, E-5 and E-6) were cancelled in 1962. It seems likely, therefore, that the purpose of the study was to address the possibility of a scaled-up version of the 77-inch Maksutov design for Gambit-3. By November of 1963 this approach was rejected leaving Kingslake's R-360 the only formula available. Referring to Kodak's limited experience in polishing aspheres, Keim makes the statement that "the present hand correcting [of refractive elements] can not be considered as true asphierizing."

Another mystery surrounds a formula sheet, the R-361, drawn up by John Villa, dated 1-27-64. The design is attributed to Tropel but the final report, containing the design data essential to a complete system analysis, went to Kodak's Optical Design Department, which was totally isolated from the ongoing classified programs. Apparently I had approved this formula for the presentation to the customer, sometime in the spring of 1964, and it was a scale-up of this formula (from 100 to a 160 inch focal length) that was converted into the "formula sample" — a full-scale breadboard of the Gambit-3 optics.

In retrospect, there were decided advantages to this design, with refractive field flatteners close to the image plane, because it did not require a hole in the stereo mirror and the asphericity of the primary could be tested by auto-collimation off of a flat.

The Breadboard

The purpose for building this rather costly breadboard, with its 43-inch diameter solid glass Pyroceram primary, was to verify the formula. But that fact got lost in the shuffle. By the time we were able to obtain test results, the lens was already in production and our evaluation of the formula became somewhat irrelevant. However, it established that the on-axis optical quality of the lens could be tested interferometrically by auto-collimation using a test flat (not shown in the picture). It also initiated test procedures as well as the computer software for wave front analysis. Converting from the double-pass to a single-pass wave front, dividing by two, provided the overall image quality in terms of the rms wave front error. Double-pass lens-film resolution measurements were also made but could not be related to a single-pass number. For

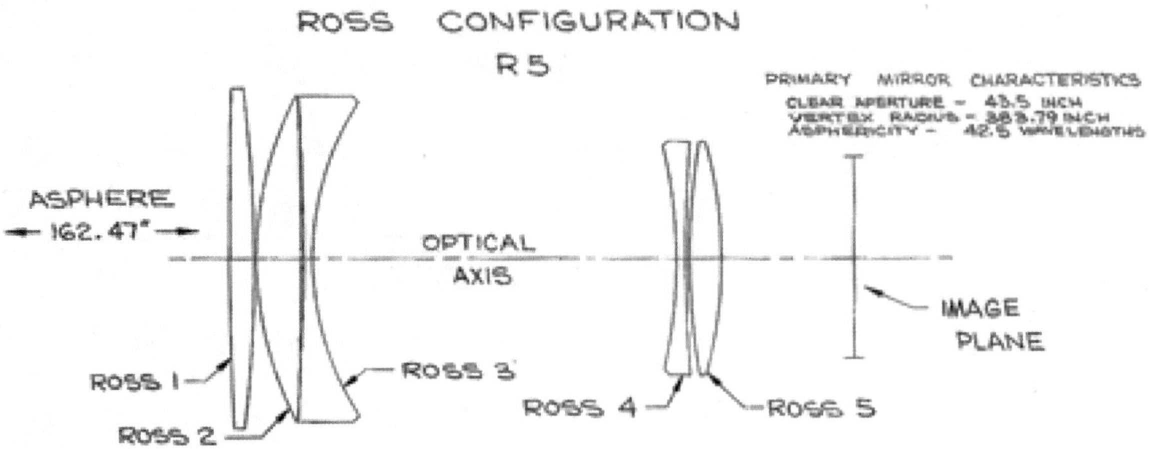

Figure 12: **R-5 Ross Corrector Configuration**
Drawing of the R-5 Ross Corrector Configuration with primary mirror characteristics. This R-5 lens formula provided a marked improvement in image quality across the field from 1970 to the end of the program.

that reason the optical quality specification was established in terms of the wave front rms error and the OQF, the optical quality factor.

Subsequent to declassification, I was surprised to learn that resolution measurements on film were being made and that the exposed flight films were sent to Bridgehead[62] for processing and then returned.

> "The Gambit program was developing one of the finest optical cameras at Lincoln Plant (locally known as K-program), some three miles west of Bridgehead. Naturally, the performance of the camera system required evaluation using the intended flight films.[63]"

Les Mitchell[64] assured me that these tests were performed at the final assembly stage to demonstrate acceptable camera operation with image motion compensation.

The R-5 Formula

So what was the formula finally arrived at for the K (or KH-8) program? In subsequent years, 1964 to 1967, optical design computer programs evolved not only at Tropel and Kodak but also in the Boston area. An independent lens designer in that area, David Grey, had developed what he called a differential correction lens design program called "the Lens II program." Apparently he was given the design parameters and submitted a formula for a 175-inch f/4 lens with four Ross corrector elements. A full analysis of this formula, referred to as the R-4,' as well as a redesigned version, with five elements, is contained in Tropel's report, dated September 19, 1968.

Only a lens designer would appreciate the subtle differences in image quality, primarily toward the edge of the field, indicated by comparing the OPD (optical path difference) curves plotted for each design. However, its important to note that, to Tropel's credit, the redesign resulted in a significant improvement. The report "...describes the performance of a redesigned version and its sensitivity to variations in partial dispersion of the glass". This was the R-5 formula used on Gambit 3.

So how good was the formula? Would the lens be limited only by diffraction? The answer to the first question would have to be: incredibly good or as good as automatic design computer programs and the country's best lens designers could make it.

The answer to the second question is: a qualified yes, but only over the central portion of the field. In terms of the size of the axial spot diagram, all rays intersected the image plane within .0004 inch or roughly half the size of the Airy Disc. A plot of the geometric response function, indicates that resolution at the 2/3 field point was 300 lines/mm. For comparison purposes, the "visual resolution" for a diffraction limited F/4 lens, that is, if one could observe the aerial image, is 427 lines/mm.[65] So even for a perfectly manufactured lens image quality would fall off toward the edge of the field.

[62]Bridgehead "The STORY OF ITS [KODAK'S] ROLE IN THE EARLY OVERHEAD PROGRAMS OF THE NATIONAL RECONNAISSANCE OFFICE," 30 December 2011

[63]Ibid. The Cue Ball operation , p. 36

[64]Les Mitchell, Program Manager SAMOS and Gambit-3 program

[65]Substituting the aperture diameter, T, in the formula, a = 4.51/T, the angular resolution is .10 seconds.

Because the formula was extremely sensitive to the partial dispersions, the second two elements had to be a matched set, based on index and dispersion data from Schott. Having the melt data to the sixth significant figure was crucial.

A Summary of the K-program (KH-8) Accomplishments

The following is a brief summary of the major innovations and new optics technology applied to the K-program.[66] In retrospect recognition should be given to the technical achievements of the earlier Gambit-1 program [KH-7], which together with performance projections made the K-program [KH-8] proposals successful. However, at that time no one fully appreciated the shear size of the primary and stereo platforms or anticipated the problems that were to be encountered in handling and fabrication.

The ultimate success of the optics development effort must be attributed to a number of technical developments that emerged at just the right time. Certainly the program could not have succeeded without the computer and the highly sophisticated programs available for structural analysis, lens design, image quality evaluation, thermal analysis, wave front data reduction etc. The primary and stereo mirror substrate development at Corning alone represented a major undertaking with a significant parallel effort to develop ultra-low coefficient of expansion material.[67] These structures were not only designed to withstand launch and the space environment but they also had to be designed to allow for valid zero-g testing on edge.

Lens Design: The lens formula called for an aspheric primary mirror for spherical correction, together with a five element Ross corrector to extend the field coverage. Automatic lens design programs were used to optimize the design and performance predictions were made with great precision using the most advanced computer programs available.

Schott in Germany provided melt data sheets with each shipment. Further analysis was done using the melt data to re-optimize the formula for matched sets of elements.

Optics Configuration: A major configuration change was implimented following the decision to perforate the stereo mirror and support the Ross corrector within the hole. This decision allowed for a simpler, more direct film path as well as a structurally improved coupling of camera to the Ross corrector housing.

Another important decision during the course of configuration studies involved the off-set between the optical and vehicle center-lines. By lowering the optical center line, beveling the stereo mirror was no longer necessary and a serious problem in polishing and testing the mirror was elimenated.

[66] This summary was prepared originally for a report to the customer sometime before I left the program in January 1975.
[67] The glass-ceramic mass produced and sold to the glass-top stove market is either the same or a derivative of this ultra-low coefficient of expansion material, ULE.

Component and System Testing: To test the primary mirror during the surface finishing operation a refractive Null Corrector was designed [at Tropel] to allow for interferometric testing by autocollimation. A two-element corrector was fabricated to state-of-the-art tolerances and on-back air bag testing of primary mirrors became routine.

Stereo-mirror test methods were developed with the mirror supported on edge and on back to simulate zero-g conditions. In both cases tests were performed by autocollimation from a master sphere with the surface figure contribution of the test sphere mathematically backed out. Good agreement between analysis and test results led to successful on back testing of the mirror. By comparing on edge and on back test data, gravitational effects in the on-edge test could be removed. This was particularly significant in generating zero-g system level test data.

Lens Interferometric Testing: Its safe to say that the invention of the laser[68] and its application to interferometric testing played a pivotal part in the development of optics for space reconnaissance. The laser was an indispensible tool, providing interferometry with a source of intense, monochromatic, coherent light. In addition to its immediate application to component testing, laser interferometry made possible remote testing of the complete optical system in the vacuum chambers, first, with the extensometer[69], and later with TV cameras to transfer interferogram images to monitors outside the chamber.

The Axicon Autocollimator: Another optical tool, developed and adapted for use in the assembly of the Ross Corrector, was the axicon autocollimator. This device, invented by J.H.McLeod, provided an extremely accurate means of aligning the corrector elements just prior to the potting operation. In essence, the conical element in the axicon projects the image of a concentrated arc source as a line in space. In the assembly of the Ross elements, the images reflected from the lens surfaces are viewed in an eyepiece and brought into coincidence with small lateral adjustments of the lens elements.

Auto-focus Electronics: Perhaps the most important development was the addition of auto-focus electronics on K program, which guaranteed in-focus imagery.

[68]Laser emission first demonstrated by Theodore Maiman in 1960

[69]The "extensometer" optically projected the interferogram, through a series of relay lenses, through a port in the top of the chamber where it could be viewed and recorded.

> Even with low-coefficient-of-expansion materials, such as Corning's ULE, invar, and thermal compensation to maintain the relative position of optical elements, the sheer size of the supporting structures made auto-focusing essential.

Testing the R-5 Formula Lenses

This brings up the interesting question: how were the lenses tested in production? First of all, it was an article of faith at Kodak that "the proof of the pudding" was in the test of a fully assembled lens. Had Perkin Elmer adhered to this policy, they could have saved themselves the embarrassment, not to mention the millions of taxpayer dollars, of the Hubble catastrophe. But that's another story.

Ideally you have large collimators to pass a parallel beam of light through the lens, simulating the "real world" single-pass imaging. I recall the six-foot diameter off-axis parabolas used at the NASA facility in Cleveland but collimators of this type were obviously impractical for the K program.[70] Interferometer tests were, therefore, performed double-pass using test-flats, which could be placed inside the vacuum chambers.

Single-pass RMS wave front error (derived from the double-pass measurement in vacuum) and the optical quality factor (OQF) were reported for every lens in production and the customer accepted the RMS wave front data to certify the optics quality.

By the 1970's interferometric testing of the mirrors, as well as the completed lens assemblies, had advanced the manufacturing technology to the point of near perfection. I remember it was a tense moment focusing the extensometer/interferometer and finally seeing the wave front of the first R-5 lens — perfectly straight fringes — and later submitting the interferogram for computer evaluation. The wave front error for the first five of the R-5 lenses was in the range .11 RMS to .08 RMS (70% to 83% Optical Quality Factor). The lenses were essentially diffraction-limited on-axis and the R-5 formula guaranteed excellent performance across the field.

This was the illusive goal of the Gambit-3 program and Kodak's crowning engineering achievement in delivering the most difficult optics to produce on an unprecedented scale.

KH-8 a Qualified Success

Soon after the last Gambit-3 launch, publishers were eager to publish authors writing about America's top-secret spy satellites.

The first book about America's secret spy satellites was published in 1986, the year I retired from Kodak. "*Deep Black*" by William Burrows was the most controversial book of the year with references to the KH-8 program and ground resolution (obviously derived from leaks to the media).

> "The first KH-8 blasted off from Vandenberg on July 29,1966, beating the first KH-7 into space by twelve days. The third generation close-look platform

[70] Testing complete lens assemblies in vacuum precluded the use of collimators.

> had a telescopic system that could produce standard photographs with resolution of six inches...[71]" [!]

At the time I knew the first statement to be quite inaccurate and, therefore, dismissed the second for what it was, an unconfirmed leak.

Richelson's book, "*America's Secret Eyes in Space*," published in 1990, was acclaimed as the definitive account of America's reconnaissance satellites, and of course I read it with great interest. The following reference reaffirms what was generally accepted — the program was an outstanding success!

> "...50 successful launches in 51 attempts...the ability of the KH-8 to return detailed photography was unsurpassed. By the later stages of the program, if not before, the two film capsules returned by each KH-8 contained imagery with six inch resolution.[72]"

Richelson's statement also reaffirms that 6-inch resolution was generally accepted in reference to "the later stages of the program" in which the R-5 lens was incorporated.

Kodakers who worked on K program look back with pride on their contributions to placing the world's finest cameras in space. They were indeed pioneers and earned their place in the history of the KEYHOLE programs, "...the most significant military development of this [the 20th] century and perhaps in all history.[73]"

More recent references to KH-8

In the June 1, 2009, issue of the "*Space Review*[74]" Dwayne Day makes even more astounding claims:

> "***The KH-8 was the highest resolution reconnaissance satellite ever built.***
> Even today, it apparently holds the record for the best reconnaissance photographs returned from orbit by any spacecraft, a combination of both a powerful camera and the ability to dramatically lower its orbit, to "swoop" in over a target at altitudes of apparently only 70 nautical miles (130 kilometers)....[and]
>
> "Although the satellite's average resolution was apparently six inches, several sources have stated that the KH-8 eventually achieved accuracy as great as 2.5 inches (6.3 centimeters) and one source has indicated that KH-8 was able to resolve golf-ball size objects" [!]

I had heard this claim, that golf-ball sized objects could be resolved, but at first discounted it as a stretch of the imagination. Was it possible to resolve golf ball size objects? How could I challenge the validity of these leaks — attributed to "several sources," presumably credible government officials?

[71] Burrows, William E., "Deep Black", p.225
[72] Richelson, Jeffrey T. "America's Secret Eyes in Space", p. 199
[73] Ibid. p. 265
[74] Dwayne A. Day, "The Space Review: Ike's gambit: The KH-8 reconnaissance satellite" 5/28/09

Prior to declassification any reference to the performance capability of KH-8 in published documents was inadmissible. Post declassification, the documents could be referenced but I could neither confirm nor deny claims of ground resolution capability. With the recent release on the Internet of the document titled, "Critical To US Security: the Gambit and Hexagon Satellite Reconnaissance Systems," the NRO has finally put to rest this intriguing question of the Gambit ground resolution capability.

Its interesting that there are several references to the Gambit-1 ground resolution in the text but only qualitative statements in reference to Gambit-3 ground resolution: "...the new lens permitted Gambit resolution to surpass even the previous best." The text went on to say, "by the time of the last Gambit-3 flight in April 1984, Gambit-3 was still producing the high quality imagery, which maintained its preeminence in technical collection". However, the chart on page 15 does list, under payload data, ground resolution of 8-10 inches (Critical to U.S. Security: The Gambit and HEXAGON Satellite Reconnaissance Systems, per website: www.nro.gov/history/csnr/gambhex/docs). This, presumably, represents an average or typical operational performance range and not best performance under ideal conditions.

To get back to the Dwayne Day (no relation) article, I should at least comment on this remarkably convincing claim with respect to the performance capability of KH-8.

First of all, I would have to say that if ground resolution at 70 nautical miles is 2.5 inches, the corresponding resolution on film would have to be 460 lines/mm, which is beyond the theoretical lens-only resolution limit for a diffraction limited KH-8 lens. As stated earlier, the aerial image resolution limit is 427 lines-mm for a diffraction-limited f/4 lens. Furthermore, I have previously given the lens-film resolution of a diffraction-limited f/4 lens and SO-243 film (resolving power 500 lines/mm), as 275 lines/mm.

The question is, how good could the lens-film resolution be with a higher resolving power film and edge enhanced development? Appendix B of the Bridgehead document[75] lists a 1020 lines/mm film as one of the films used in the Gambit Hexagon programs. However film speed would have precluded its use except on an experimental basis.

The most frequently used films are Kodak Aerial Film 3414 and the Ultra-Thin Base version, 1414, which are listed with film speed, FAI, of 15 and [edge-enhanced] resolving power of 720 lines/mm. With either of these films, lens-film resolution could theoretically be as high as 330 lines/mm, which translates to 3.5 inch ground resolution at 70 nautical miles assuming perfect operational conditions.

Considering the dynamic conditions during operation, it would have to be a rare event for this theoretical performance level to be achieved; one in which a dedicated effort is made to achieve best performance by programming a shot directly over a tri-bar resolution target with adequate time for vehicle attitude to stabilize. What that implies is that attitude rate-of-change in Yaw, Pitch, and Roll is within the extremely tight tolerance requirement. In addition there is the requirement to program image motion compensation (IMC) to an equally tight tolerance. So even on a clear day these operational variables would, in practice, degrade ground resolution by as much as a factor of 2.

[75]Bridgehead, "The Story of its (Kodak's) Role in the Early Overhead Programs of the National Reconnaissance Office." Released September 30, 2011.

Photograph 24: **GAMBIT 3 Reconnaissance Satellite Display at the Air Force Museum, Dayton, Ohio**
The KH-8 Program Successfully launched 50 satellites in 51 attempts.

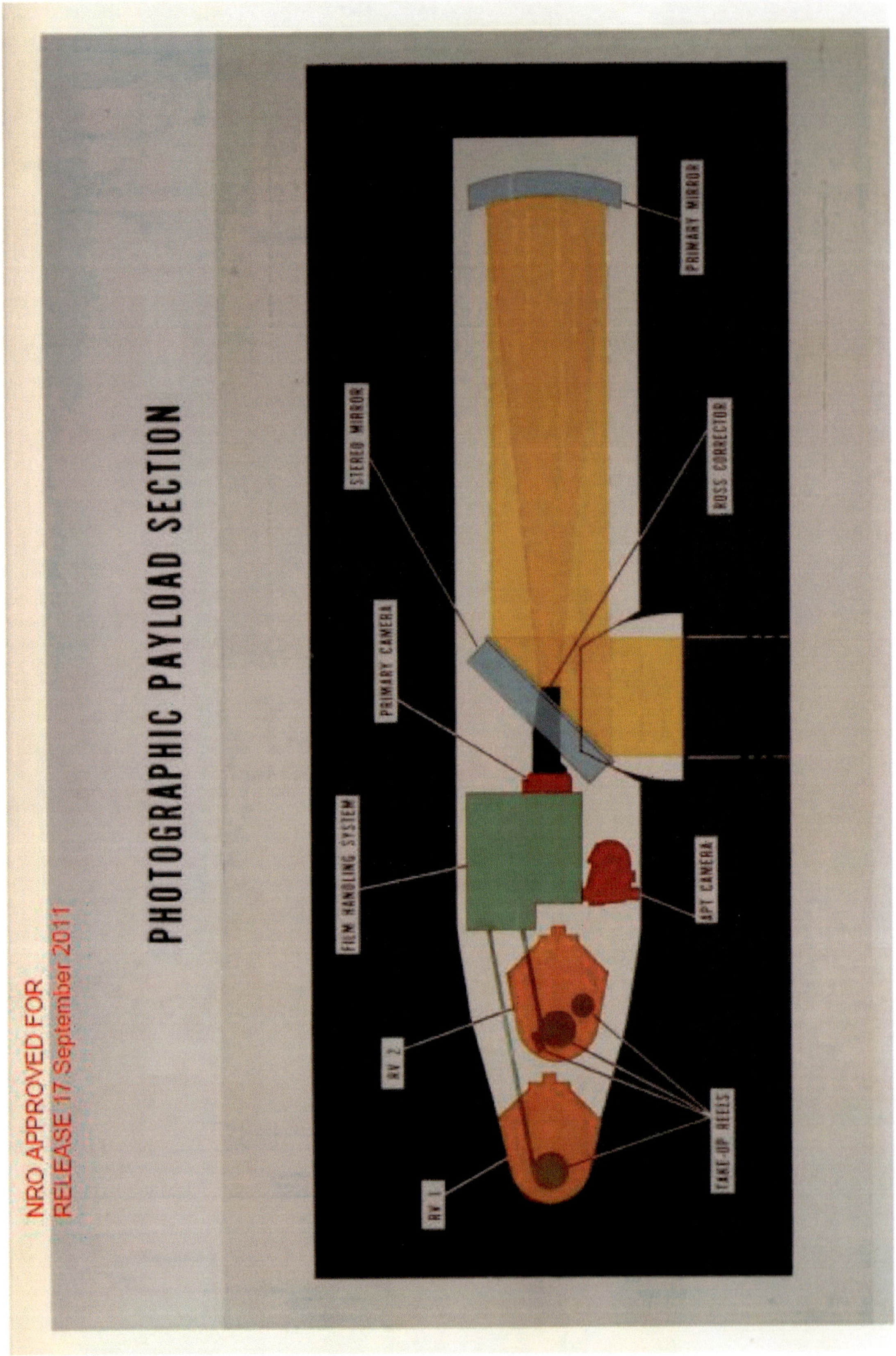

Photograph 25: **PHOTOGRAPHIC PAYLOAD SECTION**
Cross setion, released 17 September 2011. Shows Ross Corrector passing through Stereo Mirror.

The fact that "several sources stated", or rather leaked information about the performance capability of the Top Secret KH-8 program, not only reflects on the state of national security at the highest levels but it damages efforts to secure all government classified programs.

ITT CLAIMS CREDIT!

Photograph 26: **ITT Engineering for Life**
KH-8 43-inch primary mirror at the Institute of Optics.

On a tour of the new Institute of Optics building, Goergen Hall, at least two years prior to declassification, I was stunned to see the KH-8 43-inch primary mirror mounted to the wall together with a plaque that reads "ITT Engineering for Life". I was never more tempted to mention that graduates of the Institute of Optics, working at Kodak, played a pivotal part in the development of these mirrors, in fact, all the optics for the CIA's "*Eyes in Space?*" But I said nothing. Later I went back for a picture and read this inscription.

> "For more than 50 years, ITT has provided sophisticated optical technology for both ground and space borne systems and has been leading the way in a drive toward increasingly lighter-weight mirrors. The mirror on the left was built for a system that uses a combination of refracting and reflecting elements. A mirror similar to this one was used on the Hubble Space Telescope for which ITT developed a back-up mirror that is currently on display at the Smithsonian Institute in Washington."

Having acquired the Research and Engineering Division in 2006, ITT seems to have laid claim, by implication, to the development and technology behind the two mirrors, the KH-8 primary mirror and the Hubble back-up mirror. When I wrote to Chuck Spoelhof about this incredible claim, his response was:

> "...ITT has not only bought R&E but also apparently also bought its history to claim as its own. But certainly it is a distortion of history. I'm glad ITT didn't also claim credit for the film and chemistry that made these programs such a success. If there is one thing we can be very proud of, it's our success with security. We managed (and thanks go mainly to Dave Schaeffer) to keep ourselves out of the news so very few if any leaks came from us. We did not advertise our great accomplishments and felt no need to do so.[76]"

[76] A letter dated December 10, 2009

Distribution:

L. T. Askren	R. M. Price
P. M. Bogart	C. D. Rogerson
P. B. Day	J. A. Spina
A. H. Falter	D. J. Spooner
R. A. Grammer	R. W. Tietge
R. H. Hollenbeck	A. S. Wells
R. E. Keim	

19 December 1974
Sheets 1

To: Distribution

From: R. E. Keim

Subject: Personnel Reassignment

Effective January 6, 1975, Mr. Pierce B. Day will transfer to Department 195. The Optics and Mounts Group will then be the responsibility of Mr. Calvin D. Rogerson.

Mr. Day, more than anyone else, can be considered the father of the K optical system. He has seen his young "child" grow from a stubborn youngster, through a rebellious adolescent period, and finally into a mature manhood with all projections for a healthy vigorous life. Our thanks go with him along with best wishes for success in his new assignment.

REKeim:dh

R. E. Keim

R. H. Hollenbeck

R. W. Tietge

L. T. Askren

Letter 2: **K Program Cut-Backs**
This letter from R.E. Keim marked the end of my tenure in charge of the K program optics as well as a major cut back in R&E personnel. It is also a significant document in identifying a few of my colleagues and bosses including Lee Askren with whom I played tennis and repeatedly lost.

KODAK MEMO

10 March 1986

TO: Pierce Day
FROM: R. D. Sherwood
SUBJECT: Perfect Attendance

Over the years Special Programs' personnel have maintained an eviable tradition of dedication toward meeting our business obligations and commitments. You, along with 160 other people within Depts. 380-390, have contributed to this outstanding performance by maintaining perfect attendance in 1985.

I want you to know that your dependability and dedication to Program efforts, characterized by your excellent attendance record, is a valuable asset that is greatly appreciated.

I wish you good health and a challenging, rewarding year in 1986.

R. D. Sherwood

RDS:cm

***Letter 3:* Retirement Letter**

From R.D.Sherwood to the author, Subject: Perfect attendance, dated 10 March 1986. I had preserved this letter, the last official correspondence before my retirement from Kodak, not to document my attendance record but rather the words of appreciation expressed not only to myself but to the entire Special Projects staff. Sherwood was the director of the Photographic Operations Center known as BRIDGEHEAD.

10

The Hubble Telescope

The story of the Hubble telescope is well documented in the news media and elsewhere — how "after NASA expended billions of dollars and 25 years of effort, the telescope arrived in orbit with blurred vision.[77]" But little has been said of the billions previously invested by the government (from the 60's to the early 70's) to establish the technology that made it possible.

By the 70's three companies were in a position to bid on the Hubble telescope- undoubtedly the most ambitious, technically challenging project ever attempted by NASA. All three had capabilities for polishing large mirrors, made from Corning's silicon/titanium oxide mirror blanks,[78] but presumably could not refer openly to classified programs (which clearly would have given Kodak the edge).

The fact that NASA went to Perkin Elmer, the low bidder, had always been a puzzle to me.[79] Had they known anything of these programs? Perkin Elmer was a newcomer to the manufacture of optics for space.[80] Their price tag ($69.5M) was significantly lower than Kodak's (by tens of millions) because their proposal omitted a test of the fully assembled lens. In retrospect, such a test would have caught the flaw in the primary mirror and Kodak's back-up mirror, bought and paid for by NASA, might have been substituted without a hitch in the schedule.

> "A casual observer could never discern a difference between the flawed mirror and its correctly polished twin — a back-up copy built by Kodak on display at the museum.[81]"

I have Chuck Spoelhof[82] to thank for allowing me to see the report of the investigating panel. He was a member. With his permission and with apologies to those who will forever relive this episode, I had to complete the story of "The best laid plans of mice and men often gang astray.[83]"

The problem with the mirror was not in the polishing (it was polished to better than a .10 wave RMS) but in a rather odd feature of the test fixture. The reference for setting the focus of the interferometer at the center of curvature of the mirror, was the polished end of a rod which then swung out of the way for the test. Whoever designed the fixture will forever regret adding a sleeve with a hole in it, ostensibly to ensure centering of the point of focus. Unfortunately, the outer surface of the sleeve, raised about a millimeter above the reference mirror, was also polished and the procedures neglected to instruct the technician to focus through the hole.

[77] After Sputnik, Smithsonian Books, 2007

[78] Again the recorded history omits the part played by the optics group at Kodak in the development of these ultra-low coefficient of expansion [ULE] blanks

[79] Probably the determining factor in that decision was the fact that Perkin Elmer's Big Bird reconnaissance program was phased out in 1986.

[80] At the time I was not fully aware of Perkin Elmer's involvement in producing the optics for "Big Bird".

[81] After Sputnik, Smithsonian Books, p. 220

[82] Charles P. Spoelhof: From his obit December 2010 "Worked for Eastman Kodak from 1954-1986 as an optical engineer and retired as a Vice President. He was recognized as a National Reconnaissance Pioneer in 2002 by the National Reconnaissance Office for his work at Kodak."

[83] Robbie Burns

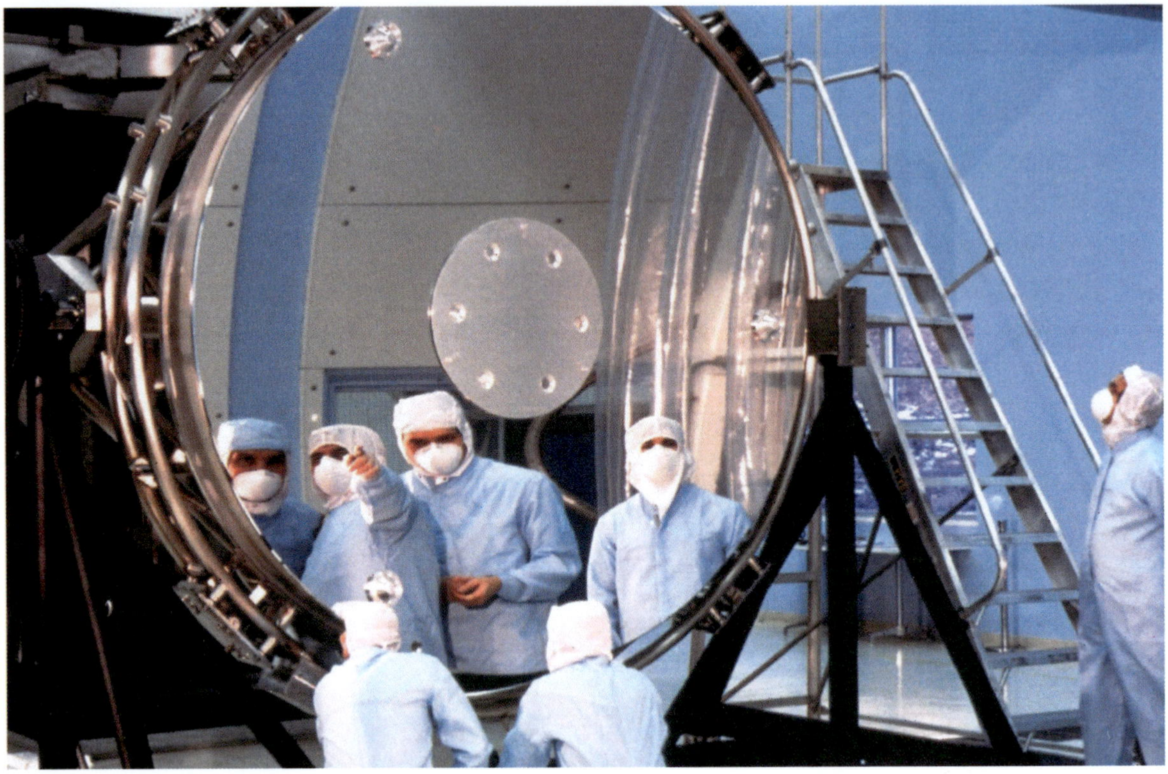

Photograph 27: **The Hubble Telescope Primary 95-Inch Mirror**
Because of a procedural flaw in testing the mirror, the surface figure introduced spherical aberration.
NASA rejected Kodak's higher bid, which included a test (omitted by Perkin Elmer) of the fully assembled lens.

Photograph 28: **Kodak Back-Up Hubble Telescope Mirror**
A similar fixture was used in polishing the Kodak mirror, but the surface figure was flawless.

The technician inadvertently focused the interferometer on the sleeve rather than the reference mirror, introducing the infamous spherical aberration.

What if: a CCD Camera!

> "It is possible to conceive of press photography in the future using CCD cameras" — James Chung, Fuji Film, 1980[84]

And what if the Hubble Telescope were turned earthward? What would be the resolution of objects on the ground?

When, in the late 60's, scientists at Bell Laboratories demonstrated a Charge-Coupled Device (CCD), astronomers weren't the only ones interested. In a recent Democrat and Chronicle article a Kodak scientist at Kodak Park claims to have invented the CCD camera. His 1978 patent supports that claim, however, the fact is the development and first application of the CCD to satellite reconnaissance cameras occurred earlier. Wikipedia gives the operational dates for the satellite reconnaissance program designated KH-11 as 1976 -1995.[85]

Since the cloak of secrecy prevented the disclosure of inventions, perhaps it was no coincidence that Steven Sasson was given the assignment to build a digital camera.

The following is the public record of Kodak's first (?) digital camera:

> "The first recorded attempt to build a digital camera was in 1975 by Steven Sasson, an engineer at Eastman Kodak. It used the then-new solid-state CCD image sensor chips developed by Fairchild Semiconductor in 1973. The camera weighed 8 lbs., recorded black and white images to a cassette tape, had a resolution of .01 megapixels (10,000 pixels) and took 23 seconds to capture its first image in December 1975. The prototype camera was a technical exercise, not intended for production."

I can remember, vividly, in the early 80's, having left Research and Engineering, that I was convinced this new digital technology would never be good enough to replace Kodak film in cameras. Wrong! But the tragedy was that that impression was prevalent at the highest levels of Kodak management. [While Kodak lagged behind Japanese companies in the low-end digital-camera sweepstakes, it was among the first to market a high-end professional SLR camera].

Just out of curiosity, I recently decided to compare the quality of the very first CCDs with the film used in reconnaissance photography. First, I was struck by the apparent size of the CCD shown opposite page 222 of the Smithsonian's illustrated history, "*After Sputnik — 50 years of the Space Age.*" The 800 x 800 pixel array (.64 megapixels) was actually quite small — only .25 inch on a side. This translates to about 62 lines/mm in units of quality commonly used for film (strange that you don't hear of "lines/mm" anymore).

Anyway at first the CCD quality did not compare favorably with film, which could resolve 700 lines/mm. This was borne out by the following quote from Burrows.[86]

[84]After Sputnik, Smithsonian Books, p. 222, 2007
[85]Wikipedia.org title=Spy satellite
[86] Burrows,William E., "Deep Black" - The Startling Truth Behind America's Top-Secret Spy Satellites, p. 235, 1986

> "Those who once dreamed that SAMOS's radio-link resolution would someday be the equal of the standard photographs that come out of the bucket-droppers have seen their vision come true. Semiofficial statements, "backgrounders," and leaks to the trade press have it that the KH-11's electro-optical resolution is inferior to that of the bucket-droppers that used standard film. While that was true at the outset of the KH-11 program, it no longer is. By 1984 the resolution of the KH-11's electro-optical system was every bit as good as that of its bucket-dropping immediate predecessor, the KH-9."

This suggests that the intelligence community considered KH-11 a success if the resolution was as good as the resolution of its immediate predecessor, KH-9 (1.5 foot ground resolution). And it was the concept embodied in SAMOS of real-time reconnaissance that was vindicated.

Assuming that the CCD quality improved in parallel with that of commercially available sensors, the full potential of the superb KH-11 optics would not have been realized for another decade or the mid-90's. Only then could it have approached the six-inch (0.15 m) ground resolution of KH-8.

Reflecting on NASA and Man in Space

Although my connection with optics in space ended in January, 1975, I continued to follow America's obsession with space, which was driven by the need to prove its technological superiority over the Russians, to put a man on the moon. That summer of 1969, July 20th, we were camping in the Grand Tetons, crowded around the TV at the Jackson Hole Lodge to see Neil Armstrong's first steps on the lunar surface.

And in October 2008, I happened to switch to the NASA channel in time to see the celebration of the 50th anniversary of NASA, taking place at what was originally the Lewis Flight Propulsion Lab at Cleveland. As I sat mesmerized by this historic gathering of the astronauts, the film highlights of Gemini and Apollo, John Glenn's address, stressing the urgent need to train scientists and engineers, I thought of how I might have had a career at NASA.

I had always considered the value of man in space questionable for reconnaissance, and in fact wrote a report to that effect, but never thought of the politics or the urgency of beating the Russians to the moon. But the idea of a large space telescope to peer into the heavens, that was something different. I knew the technology would benefit science and astronomers around the world. My fascination with the Hubble Telescope, when it was launched in June of 1990, was not only in the marvel of this ultimate achievement in optics but in the fact that it revealed to the world what had been veiled in secrecy — that optics of this magnitude had long since been deployed in space.

It is hoped that as government agencies see fit to review and re-assess policies governing the classification of these programs,[87] consideration might be given to the current problem of recruiting qualified physicists and engineers. Security has turned them away because it has removed the history, the feedback from one generation to the next.

[87] The Gambit programs were declassified September 17, 2011.

On January 1, 1975, my tenure with Research and Engineering came to an end after 16 of the most challenging and fascinating years of my career in optics. Keim's letter (Letter 2) documents my transfer to what was then the Elmgrove Plant, Department 95. Although Keim was kind to suggest that I was the father of the K program optical system, he may have set me up for a paternity suit.

11

Transferred To Copy Products
1975 - 1985

In 1975 there was a general exodus from Research and Engineering (R&E) back to commercial engineering jobs, many of which were on the copier program which had just begun to take shape. I was sent back to the "real world" with a memorable party, at which I was chided, chastised and characterized by colleagues who I had thought to be friends. The door prize was a 25-dollar bond. Actually to this day the bond of mutual respect has lasted as we get together at monthly R&E luncheons.

The return to the commercial world at the new Elmgrove Plant was a dramatic change from the disciplined, goal-driven programs within the research and engineering division. The main floor at Elmgrove was a huge area, partitioned off into office spaces just big enough for one or two desks. Most of the camera and projector design and engineering functions had been transferred from Camera Works as well as all the parts manufacturing and assembly operations. It was a very busy place, but for months it felt like the Air Force staging area at Fort Dix waiting for a real assignment.

To revert back in time to 1958, my last recollection of the commercial world, Kodak engineers were working on a revolutionary new camera, one that you could slip into your coat pocket. The artist's concept drawings had everyone enthused and the mechanical design progressed, no doubt, in parallel with the lens design. Kingslake, came up with a wide-angle lens to cover the sixteen millimeter format but manufacturing tolerances were extremely tight. I can remember going over the formula sample test data with Mike Bernhardt and saying they could never reproduce this quality in production. And they didn't. Customers were disappointed.

It was a great idea, even if the execution left something to be desired. Just as digital technology transformed "America's Secret Eyes in Space," digital technology transformed the pocket camera, not to mention the entire photographic industry. Even cell phones can now take pictures which are then sent via the internet to friends and family. Our grandson, Ryan, was visiting at our cottage in Pennsylvania the day his sister, Allison, down in Texas, gave birth to our first great grandson (Gregory Allen Jaks). He was able to make this historic announcement with a picture on his cell phone — pretty cool.

The Evolution of the Pocket Camera — from Film to Megapixel Chips

While I had absolutely nothing to do with the evolution of the pocket camera, I was fascinated with its history, driven by that idea of a small camera, then the idea of instant images, instantly viewed and shared over the internet. It was a transfer of government initiated and funded digital technology to the commercial world. According to Wikipedia "The first recorded attempt to build a digital camera was in 1975 by Steven Sasson, an engineer at Eastman Kodak. It used the then-new solid-state CCD image sensor chips developed by Fairchild Semiconductor in

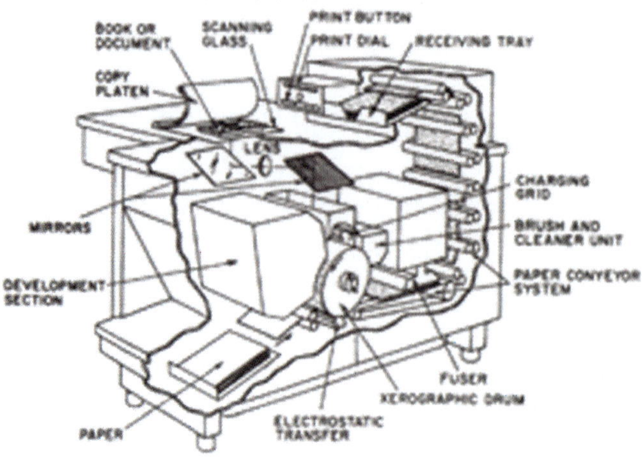

***Figure 13:* The Xerox 914 Schematic**
Schematic diagram of the Xerox 914 copier (1960). It took Carlson 15 years to establish the basic principles of electrophotography.

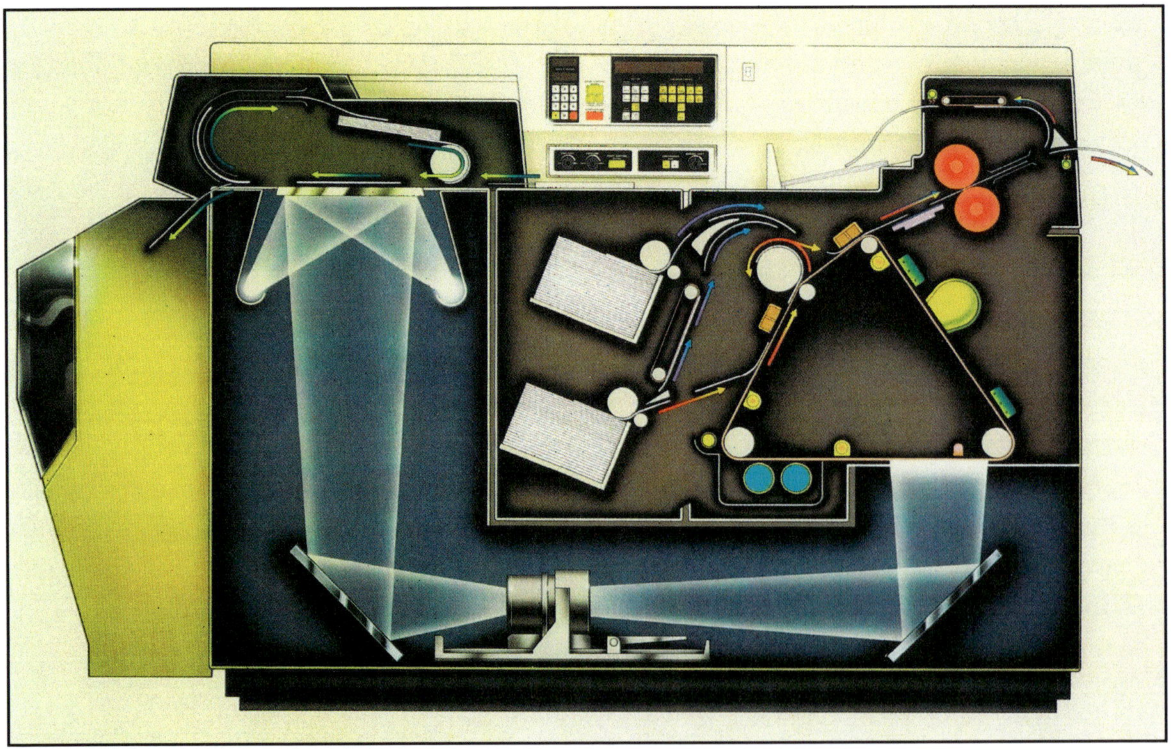

***Photograph 29:* Cross-Section of the Kodak Ektaprint 100 Copier/Duplicator**
Cross-section of the Kodak Ektaprint 100 copier-duplicator (1975). In negotiations with Xerox, Kodak attorneys used the re-circulating feeder as a bargaining chip.

1973. The camera weighed 8 lbs., recorded black and white images to a cassette tape, had a resolution of .01 megapixels (10,000 pixels) and took 23 seconds to capture its first image in December 1975. The prototype camera was a technical exercise, not intended for production.

By 1981, Sony had demonstrated a hand-held electric camera but it was an analog camera that recorded television-like images on a 2x2 inch floppy disc. At the 1984 Olympics Canon demonstrated one of these analog cameras and in 1989 images were transmitted over telephone lines from China during the Tienemin Square protests! This capability to transmit images without a satellite link was useful during the Gulf War in 1991, but image quality for news paper graphics was poor.

This reference to the transmission of imagery via satellite, of course, implied the existence of real-time reconnaissance, which was crucial to military operations in the Gulf. As a once active systems engineer, when I see a reference to image quality I automatically have to think of the contributors to poor quality, for example, telephone line losses and the resolution of the imaging sensor. How good were they in 1991?

To get back to the history of digital cameras, and where was Kodak, its interesting to note that Kodak's DCS-100 professional SLR camera was introduced in 1991! It used a 1.3 megapixel sensor, and was priced at $13,000. So it's quite possible that Kodak had cameras on the ground as well as in satellites during the Gulf War. In fact, these Kodak professional cameras may have been the first true digital cameras to use solid-state CCD sensors. Kodak also introduced the first digital camera with compact flash — the Kodak DC-25 in 1996.

Obviously, it was not in Kodak's best interest to enter the small digital camera field as long as it could still sell cameras and film in world markets, China having been the last to convert to digital.

From Verifax to Xerography

When I arrived on the copier program during the summer of 1975, the first engineering models were being tested. So some time prior to that Kodak had committed resources to the program — recognizing a real potential for revenues. But it was at least ten years behind the times.

In 1947 Haloid, a small company about a mile west of Hawk-Eye on Driving Park, had acquired a license from Battelle and produced for the first time a semi-automated xerographic machine, called the Copyflo in 1955. That same year Ricoh introduced the RiCopy 101 Diazo copier and in 1960 Xerox started to produce the Xerox 914 copier, a fully automatic desk-size copier. By 1975 small office copiers were soon on the market by Panasonic, Yoshiba, Sharp, Konica and Canon.

So where was Kodak? Actually Kodak was doing very nicely, thoughout the 50's and 60's, with Verifax desktop copiers for which Kodak could sell coated paper and developer. Verifax was a wet process in which the item to be copied is placed face down against a sheet of photo paper and exposed through the back transferring the image. The photo paper is then passed through developer liquid in the base, squeegeed and dried. It was rather messy and slow but that was the state of the technology and people accepted it- that is, until someone by the name of Chester

Carlson, working in a patent office in New York, decided there had to be a better way. They say necessity is the mother of invention, but in his case, it was frustration.

Having been somewhat involved, some 40 years after Carlson first applied for a patent, in the subsequent embodiments, Kodak copiers, I felt obligated to acknowledge not only a great inventor but also the extraordinary evolution of the invention. It's a fascinating story extracted from Wikipedia.[88]

The most tedious and time-consuming part of his job was making multiple copies of documents for the U.S. Patent Office, using a Photostat process. It was his job in the patent department of Bell Telephone Laboratories that stimulated thoughts of inventing a better way of making copies. Months of research at the New York Public Library led to his discovery of photoconductivity. He recognized the potential of substances such as zinc, selenium, arsenic and tellurium, which are "photoconductors," that is, they are insulators and hold a charge in the dark but are conductors when exposed to light. If the charge on the photoconductor was proportional to the light and dark areas in the exposed image then, with electro-statically charged powders brushed on to the photoconductor, that image could be transferred to paper. That was the basic idea and the basis for his first patent, which he applied for in October 18, 1937. The rest is history, except, that to convince anybody that it would work he needed proof — a demonstration.

Almost exactly a year later, October 20, 1938, he was able to demonstrate the process. He had proof— tangible evidence that the process would work. First he printed the words, 10-22-38Astoria, in India ink on a glass slide. This he placed face down on a photoconductive zinc plate, which he then exposed to light. After removing the glass slide he pressed a sheet of wax paper against the zinc photoconductor transferring the image to the wax paper. Finally by heating the wax paper, the wax melted off and he had his first near perfect copy. "EURICA"— the word, according to Webster, derived from the Greek word "heureka" meaning 'to have found,' an exclamation of triumph on a discovery — might have been the most likely expression of the emotion of that moment — the moment of truth.

> "It took Carlson 15 years to establish the basic principles of electro-photography. The road to success... had been long and filled with failure."

Failure to find a sponsor very nearly put an end to his quest to commercialize xerography. He was turned down both by GE and IBM. They said there was no market for such a device. The Navy was interested but had other priorities. Finally, he was able to convince the directors at Battelle Memorial Institute in Columbus, Ohio, to commit a substantial investment in the development of the process. By 1946 Carlson succeeded in establishing the feasibility of a commercial product and contracts between Battelle and Haloid Company were negotiated. In 1955 Haloid first produced an automated xerographic machine, the Copyflo duplicator, which was succeeded by the Xerox 914 Copier (7 copies/ minute) introduced in 1960. Haloid Xerox became Xerox in 1961.

[88] Wikipedia.org/wiki/Chester-Carlson

The Kodak EKTAPRINT 100 Copier-Duplicator

In April of 1975, Kodak entered the high volume electrostatic copier market with the Kodak EKTAPRINT Copier Duplicator. At 4200 copies per hour, it was 10x faster than the first Xerox 914 copier introduced in 1960 and competitive with the current high volume copiers in the high-end segment of the copier industry. It could handle originals and copy paper up to the standard 8 1/2 x 14 inch legal size.

At the time, it was an ambitious program given Kodak's somewhat limited capabilities in the commercial side of its business. I was told that the first model presented at trade shows was plagued with problems that demonstrated, if nothing else, that it was not even close to a marketable product. The company was faced with major challenges that were both technical and organizational.

With workforce reductions in the Research and Engineering Division, Kodak suddenly had a large group of engineers available to fit into the already existing Copy Products organization. In fact, a total reorganization was required with most of the supervisory positions going to R&E people. It was back to the drawing board with new ground rules and disciplines brought over from the government programs. It was necessary to start from scratch by defining requirements for every component and subsystem.

At some point, later that year, I found myself in charge of a small group responsible for the optics including the document illumination. We were on the 4th floor of Hawk-Eye right above the tool engineering department, my last contact with Al Fultz and the commercial world. It would be difficult to recount the specific challenges encountered except that there were cost-benefit trade-offs that I was unaccustomed to in government work. I was soon to encounter a whole new breed of engineer, the optics-manufacturing engineer. They seemed to have irrefutable reasons why it couldn't be done in production. I still remember the heated debates with one of these optics- manufacturing engineers who thought I was totally out of touch with reality. In all fairness, however, the lens specifications were significantly more detailed than those previously applied to commercial lenses.

In a Special Projects Development Memorandum, "Ektaprint Lens Quality Ranges and Development Of a Useful Conversion Chart," I gave the specification as follows:

> "Modulation Transfer Function (MTF) at 6 cycles per millimeter to be 40% minimum over 90% of the full field (11 degrees semi-field angle) through a depth of 1 mm".

First of all, in 1975, the basic concepts of Sine-wave Response and Modulation Transfer Functions (MTF) were new to commercial optics manufacturing. One of my first Special Projects reports (SFR-75-0015, dated October 23, 1975) presented a cursory description of the basic concepts of sine wave response as background information "for those less familiar with this well-established[89] mathematical modeling technique."

[89] R.L.Lamberts, G.C. Higgins, and R.N.Wolfe, "Measurement and Analysis of the Distribution of Energy in Optical Images," J. Opt. Sec. Am., 48: 487-490 (1958)

***Photograph 30:* Kodak Ektaprint 100 Brochure**
Cover of Kodak's Ektaprint 100 copier-duplicator advertising brochure. This was Kodak's entry into the high volume, high-end segment of the copier industry.

In the Research and Engineering Division, of course, sine-wave response functions were developed and used extensively in predicting lens-film performance. In fact the optical quality specification for those lenses was stated in terms of MTF or the equivalent Optical Quality Factor (OQF). The irony of it was that the optical manufacturing engineers knew nothing of the lenses ten times the size, being built in the R and E Division of Kodak, to a requirement of 70% OQF.

The EKTAPRINT 150 Copier Duplicator

With the EKTAPRINT 100, Kodak at least had its foot in the door and established a reputation for excellent copy quality. Within a year Kodak added the larger and significantly more versatile EKTAPRINT 150 Copier-Duplicator plus two feeders and a finisher, which provided high productivity and flexibility. With the feeder and finisher accessories, the copies could be collated, jogged, stapled and stacked at the same time. The EKTAPRINT 250 duplicators increased speed to 5500 originals per hour with built-in microcomputer-controlled automatic two-sided copying.

Patents on the Kodak EKTAPRINT 100 and 150 re-circulating feeder[90] gave Kodak an important bargaining chip in its negotiations with Xerox. Kodak also held patents on organic photoconductors, which could be coated onto a flexible film base. A flexible photoconductive film belt provided great flexibility in the placement of the various elements of the process. Needless to say Kodak's patent attorneys had their work cut out for them during those early years of the program. Presumably some of the original Xerox patents had expired which gave Xerox exclusive rights in the market place. Eventually the attorneys succeeded in arriving at an accommodation with Xerox. And the market accommodated both companies from 1975 until Kodak sold Copy Products in 1985.

Copier Lenses

Figure 15's sequence of graphs that show the development and use of the geometric mean MTF to describe copier-lens performance. The fourth graph gives the geometric mean MTF for three existing lenses and boldly sets the design goal.

Figure 16's sequence of graphs that show the complexity of representing the copier-lens thru-focus performance. The third graph presents a typical plot of Model C test data and indicates the requirement in terms of a 1.2 mm depth-of focus corridor.

In 1975, when I assumed responsibility for the copier optics, the lens and the basic configuration for Model C had already been established. In other words, I inherited an existing lens and some test data around which the specification was written. Discussions with the optics-manufacturing group centered around the manufacturing quality of the lens and the more complex specifications for the Model D lens before and after assembly. A depth-of-focus corridor had to be defined based on MTF measurements from the center to the corners of the format.

[90]Designed and presumably patented by Gary Gustafson: Hawk-Eye 1975

KODAK APPARATUS DIVISION — OPTICAL ENGINEERING DEPARTMENT

FORMULA 72-L-257
OBJ. DIST. 0.0
SEMI-FIELD 12.00 DEG. MAX.
SEMI-DIAG. 209.80
FORMAT 8.5 X 14 INCHES

FRONT ELEMENT FOCUSSING
PRINTER LENS
MORE COMPACT THAN L-256
PRINCIPAL WAVELENGTH 600
484-690 ACHROMATISM

OBJECT-IMAGE = 1986.09
TO CHANGE MAG., ENTIRE LENS IS MOVED WHILE ELEMENT COMPRISED OF SURFACES 3 & 4 IS INDEPENDENTLY SHIFTED TO ADJUST FOCAL LENGTH.

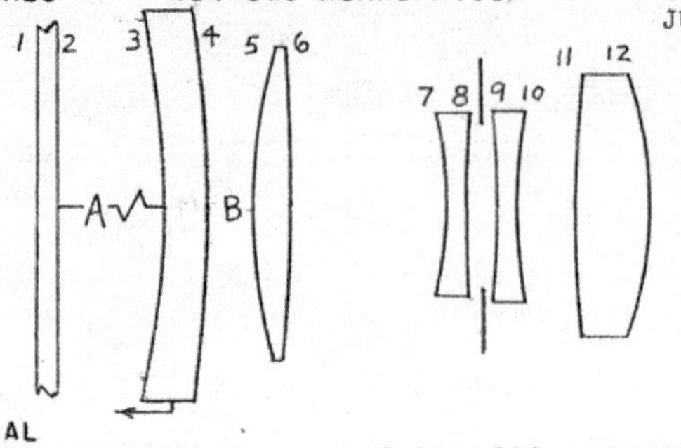

SURF.	CLEAR APER.	AXIAL BEAM	RADIUS	THICKNESS	GLASS	SPECIAL NOTES
1	514.40	0.00	PLANO	6.350	523-586	
2	512.83	0.29	PLANO	A		
3	127.14	59.91	-291.813	14.000	620-603	
4	126.20	61.67	-469.662	B		
5	103.85	65.42	193.099	13.500	620-603	
6	102.65	64.75	-411.489	52.173		
7	59.56	53.89	-133.468	7.000	617-366	
8	56.43	54.17	631.552	5.104		
	54.65	54.65	DIAPHRAGM	5.266		
9	57.62	55.19	-536.970	7.000	548-458	SCHOTT LLF-1
10	62.27	56.29	198.965	19.498		
11	80.93	62.62	413.397	25.000	620-603	
12	88.33	65.55	-135.421			

MAG.	E.F.	B.F.	F.F.	REL. APER.	EFF. APER.	IMAGE DIST.	BEST FOCUS
1.01X	497.41	451.04	375.29	7.27	14.61	953.42	-1.33
.767X	487.59	448.07	342.86	7.20	12.72	822.05	-0.90
.642X	470.29	442.83	285.73	7.07	11.61	744.76	-0.86

MAG.	A	B	LENS LENGTHS (3-12)	N.P. LOC.	N.P. DIAM.	X.P. LOC.	X.P. DIAM.
1.01X	863.60	15.512	164.053	110.87	68.42	-57.88	70.00
.767X	974.39	35.660	184.200	127.04	67.43	-57.88	70.00
.642X	1014.09	73.206	221.746	155.99	65.69	-57.88	70.00

ALL DIMENSIONS IN MM
DESIGNER P.L.RUBEN
ASSIGNMENT 3059-1
SCALE .4 X 12-05-73

***Formula 6:* Kodak Ektaprint 100 Lens Design**
Lens design for the Kodak Ektaprint 100 series copier-duplicator. Copies could be made at 1x, or reduced either at .767x or .642x. This was accomplished by moving the entire lens assembly and at the same time varying the air space between the meniscus and the other four elements.

Special Products Development Memorandum

Ektaprint Lens Quality Ranges and Development Of a Useful Conversion Chart

P.B.Day

Abstract:

Ektaprint lens Quality Ranges (Model C and Model D) in terms of MTF are derived from existing depth-of-focus data and compared with recent measurements on a Xerox 8200 lens, A full discussion of various lens quality criteria is presented in developing the basis for a useful conversion chart.

Unfortunately, the report was not dated. But I believe it was prepared shortly after Copy Products moved to the new building on Ridge Road, in 1975. The engineering group was officially named "Special Products Development" under Jim Maher. Work had begun on the next generation high-volume copier-duplicator that was referred to as the SP-4000 (later to become the Kodak Ektaprint 250 series). As explained later, the lens was simply the SP-4000 lens. In a footnote I refer to an earlier report documenting measurements comparing Kodak lenses with the Xerox 8200 lens. This was evidence that we paid close attention to the competition and later this evidence proved vital in a major flap over a significant difference in copy quality.

Just to indicate "the complexities that arise in attempting to describe and specify lens quality," I have included two charts from the report (Figure 15 and 16). Figure 15(a) shows the geometric mean MTF vs Field Position in relation to the radial and tangential MTF. Figures 15 (b) and (c) show the geometric mean MTF for the Model C and the SP-4000 lenses; and (d) sets the design goal. Figure 16 (b) shows how the thru-focus MTF, both tangential and radial, varies with field position (Model D design data); Figure 16 (a) is a plot of the geometric mean MTF vs field position and distance from best focus; and Figure 16 (c) shows a plot of actual test data for the Model C lens and the depth-of-focus corridor for that lens at .767 magnification.

The report goes on to justify the use of the geometric mean MTF and the shift in spatial frequency for the Model D and SP-4000 lenses:

> "Copy quality experiments showed that a significantly higher correlation was obtained with the lens MTF measured at 4 cycles/mm than at 6 cycles/mm. The MTF specifications were then changed to 60% at 4 cycles/mm and the D.O.F. (depth-of-focus) at 1x increased to 2 mm based on the distribution of reported D.O.F. values for the engineering model lenses."

It's interesting that the report in which these specifications are spelled out refers only to the Model C and Model D lenses with no mention of lens formulas. This was typical of the way lenses became identified with the product or a general end-use of the lens as, for example, 3-inch Cine Ektar, a 6-inch aerial camera lens or a Television Ektar Lens. It reflected to some degree the isolation of Kingslake's lens design department from the departments responsible for manufacturing the lens. After a review of the formula sample test data, there was very little contact with the manufacturing process. Apparently this had all changed when Bill Price[91] had become head of the department in the early 70's after Kingslake retired. The first copier lens was a Paul Ruben design, the L-257, dated 12-05-73, issued by the Optical Engineering Department.

[91] Price, William P., MS., 1949

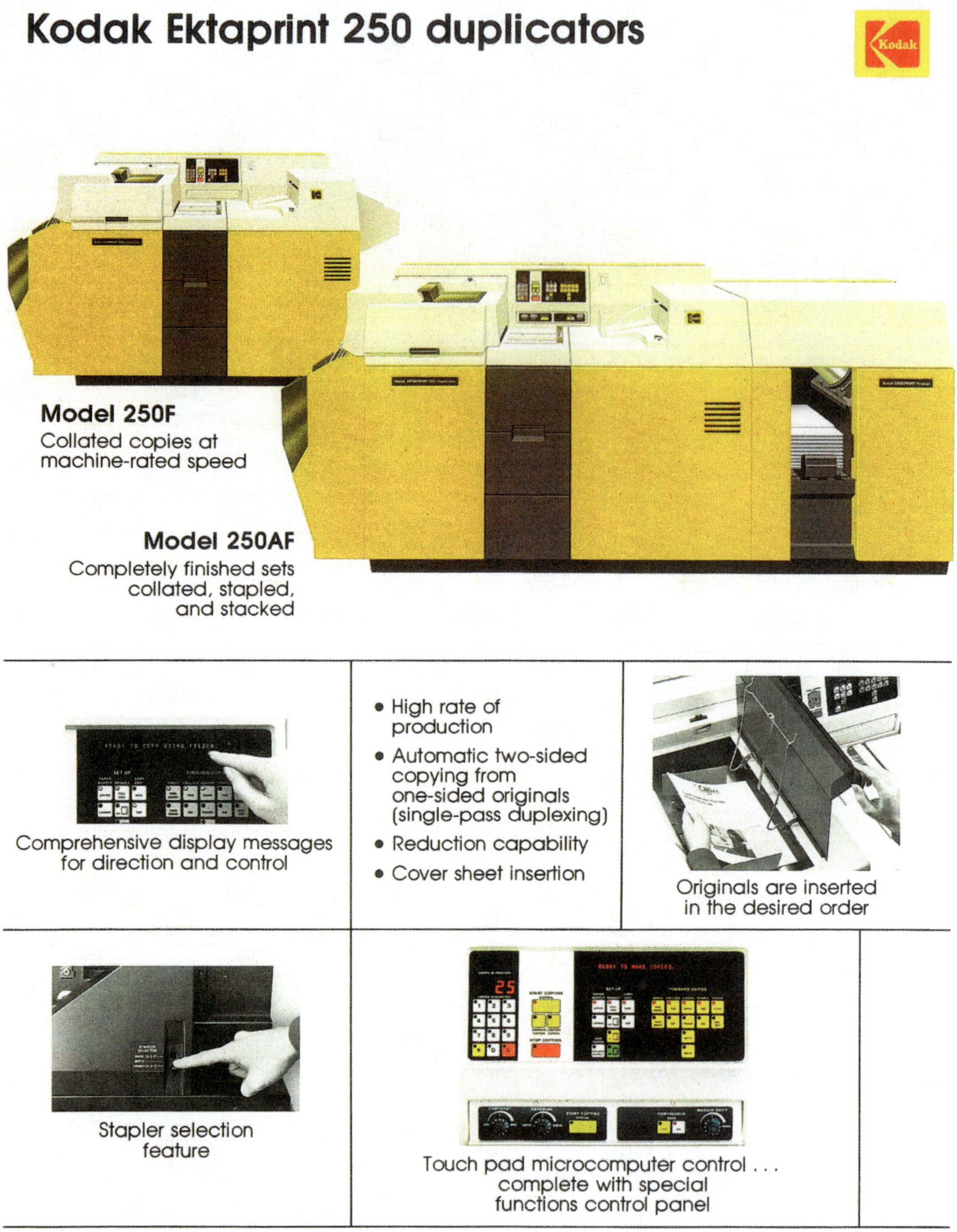

Photograph 31: **Kodak Ektaprint 250 Duplicator**
The Kodak Ektaprint 250 duplicator sold well in Washington's paper mill, providing the latest in high-tech features: high rate of production, automatic two-sided copying and touch pad microcomputer control complete with special functions control panel.

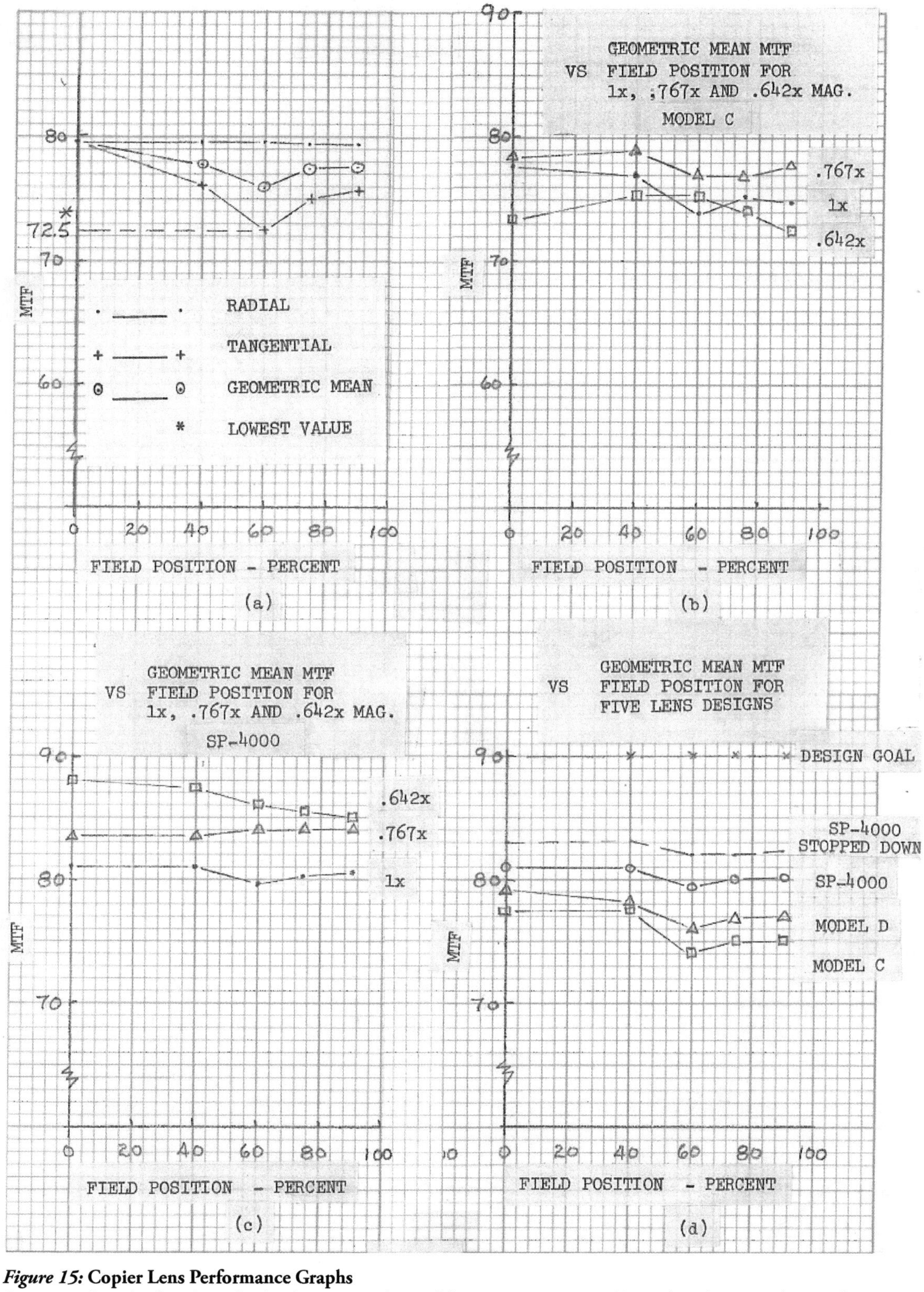

Figure 15: **Copier Lens Performance Graphs**
Sequence of graphs that show the development and use of the geometric mean MTF to describe copier-lens performance. The fourth graph gives the geometric mean MTF for three existing lenses and boldly sets the design goal.

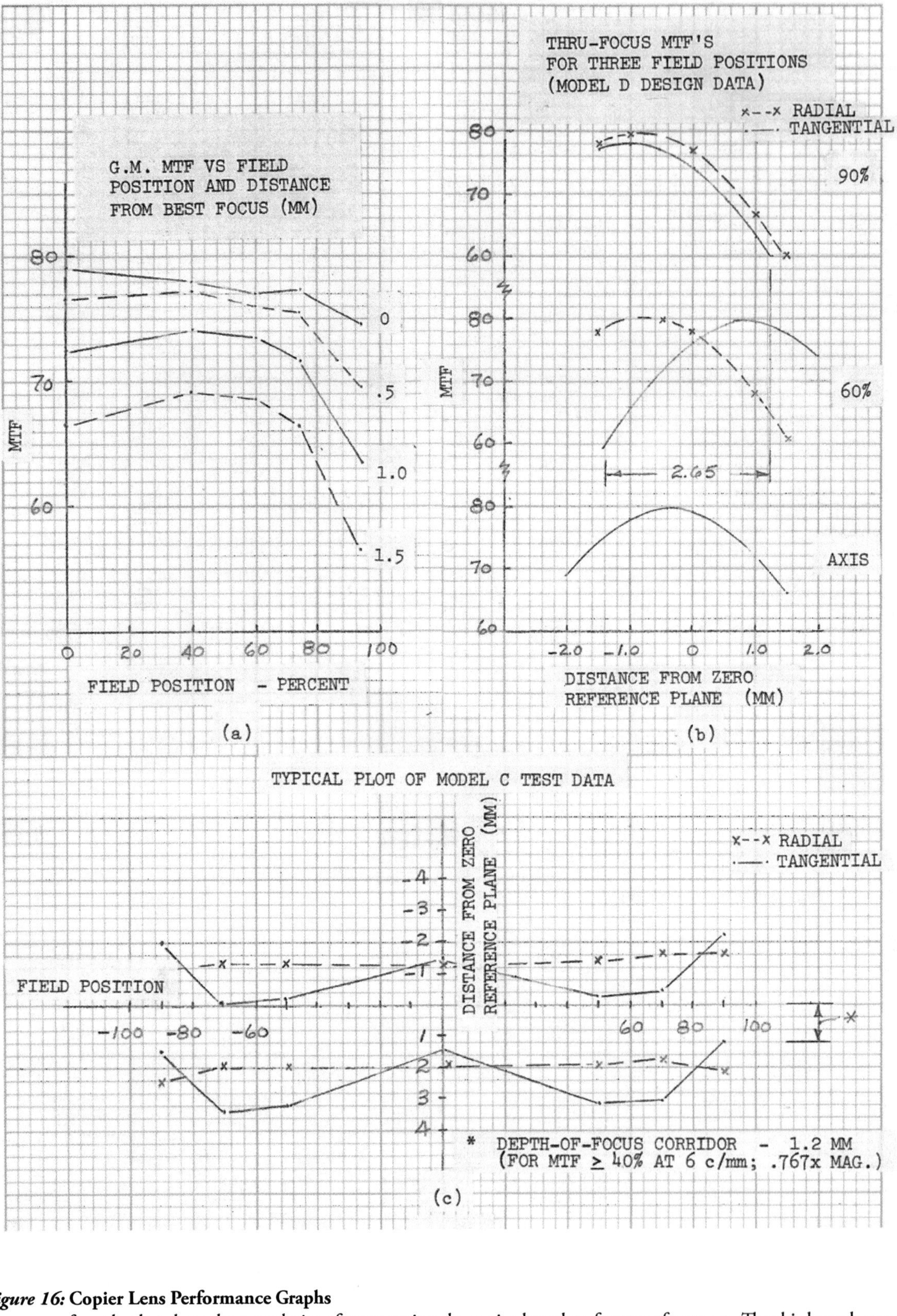

Figure 16: **Copier Lens Performance Graphs**

Sequence of graphs that show the complexity of representing the copier-lens thru-focus performance. The third graph presents a typical plot of Model C test data and indicates the requirement in terms of a 1.2 mm depth-of focus corridor.

Clearly, I was unaware of the implications of this change in title and for years in research and engineering I had distanced myself from what I always knew as the lens design department.

Perhaps it was the fact that Building 951 was somewhat remote from the rest of Kodak. Optical Engineering had moved from Hawk-Eye to the Elmgrove Plant. With all due respect to Bill Price as head of the lens design department, we simply didn't communicate. In fact I'm not sure when he became aware of the fact that I had initiated a lens design contract with Tropel. No doubt it was when I published a report titled "History of Lens Design Performance Improvement" which showed the progression of improvement that the Tropel lens design was better than the in-house designs. I understand now why Bill Price was upset, probably furious, as I look back over the dates and the considerable body of lens design effort that went into the Copy Products lenses just prior to or even parallel to the Tropel contract.

The design for the Model C copier was the L-257A formula with the L-257B a slight improvement for the Model D. There were further improvements in the U-182, U-183 series. These formulas, however, were never passed on to me in Special Project Development.

Tropel Lens Design Study

As indicated in Figure 1 (d) of my report, I had set what I believed to be an attainable design goal. Not that I thought less of Kodak's lens designers, but I felt there was room for improvement and it was important to have an independent lens design study in which I could specify a performance requirement close to the diffraction limit. That was the goal, at least, and I would be in the loop as far as the design was concerned. It never occurred to me that with Maher's approval I needed to notify the Optical Engineering Department. I wanted the final report, which would include a complete analysis of the formula with design trade-offs, to come directly to me.

George Schnabel, at Tropel, worked for several months on a design that did achieve the required performance at 1x magnification but was unable to maintain that quality level at .64x magnification, without multiple air space adjustments. When I submitted the Tropel design to Optical Engineering, Bill Price was justifiably upset and I believe it was Bob Sterrett[92] who had to make amends by inviting Bill over to Building 951 for a tour.

Shortly thereafter, I receive a copy of a report, dated June 24, 1980, by Lee Estelle, a highly regarded Kodak lens designer, outlining a redesign task "in an effort to make it manufacturable." This he accomplished by adjusting air spaces and lens thicknesses but then he added a large meniscus element and imposed a constraint that radically changed the Tropel approach. By adding the meniscus he was able to go back to adjusting only one space together with moving the total assembly. As a result the image quality turned out to be "totally unacceptable" at .64x magnification.

[92]Robert Sterrett succeeded Jim Maher as head of the Special Projects Development

However, he does admit that the Tropel design improved secondary spectrum over the U-182 and concludes with the statement:

> "To achieve the highest quality level possible for lens designs of this type, I would agree with the Tropel conclusions, namely that more than one moving element is necessary."

The Proof of the Pudding

Apparently, I did consult with Price before taking the next step in the saga of the Tropel lens. In my June 5, 1981 letter to Maher,[93] I had also given John Buzawa[94] at Tropel the Lee Estelle design modifications for improved manufacturability. Attached to the letter was the Purchase Order to Tropel to fabricate two unit magnification lenses based on the lens design study. So at that point the lens was still viable for inclusion in the Ektaprint 250 copiers in spite of the lower performance at reduced magnification. Evidently it was a trade-off that could only be evaluated with prototype lenses installed in the engineering model machines.

MTF measurements on one of the Tropel lenses showed actual MTF data points falling almost exactly on the design MTF curve. One could hardly ask for better agreement and proof that the projected goal of 90 percent MTF across the field was attainable. But still, Optical Engineering claimed the cost to manufacture the lens was prohibitive. In the end, Lee Estelle was able to match the Tropel design with a six element M type formula. At least that's what I was told. I seem to have acquired the formula but no design data. I was out of the loop.

Ironically, I was the one confronted with the charge that Xerox had better optics and that's why Xerox copiers produced sharper text. I had to convince management that it was something about their process that gave that edge sharpening effect. It helped to have the MTF measurements on the Xerox 8200 lens — obviously removed surreptitiously from one of their machines.

[93]As an artifact of Kodak history, I thought it rather interesting that the letter is on Kodak Park stationary celebrating Kodak's centennial- "A 100-year Start on tomorrow". Sadly, Kodak Park's tomorrows came to an end sooner than anyone anticipated.

[94]Buzzawa, M. John, BS, 1952; MS, 1954

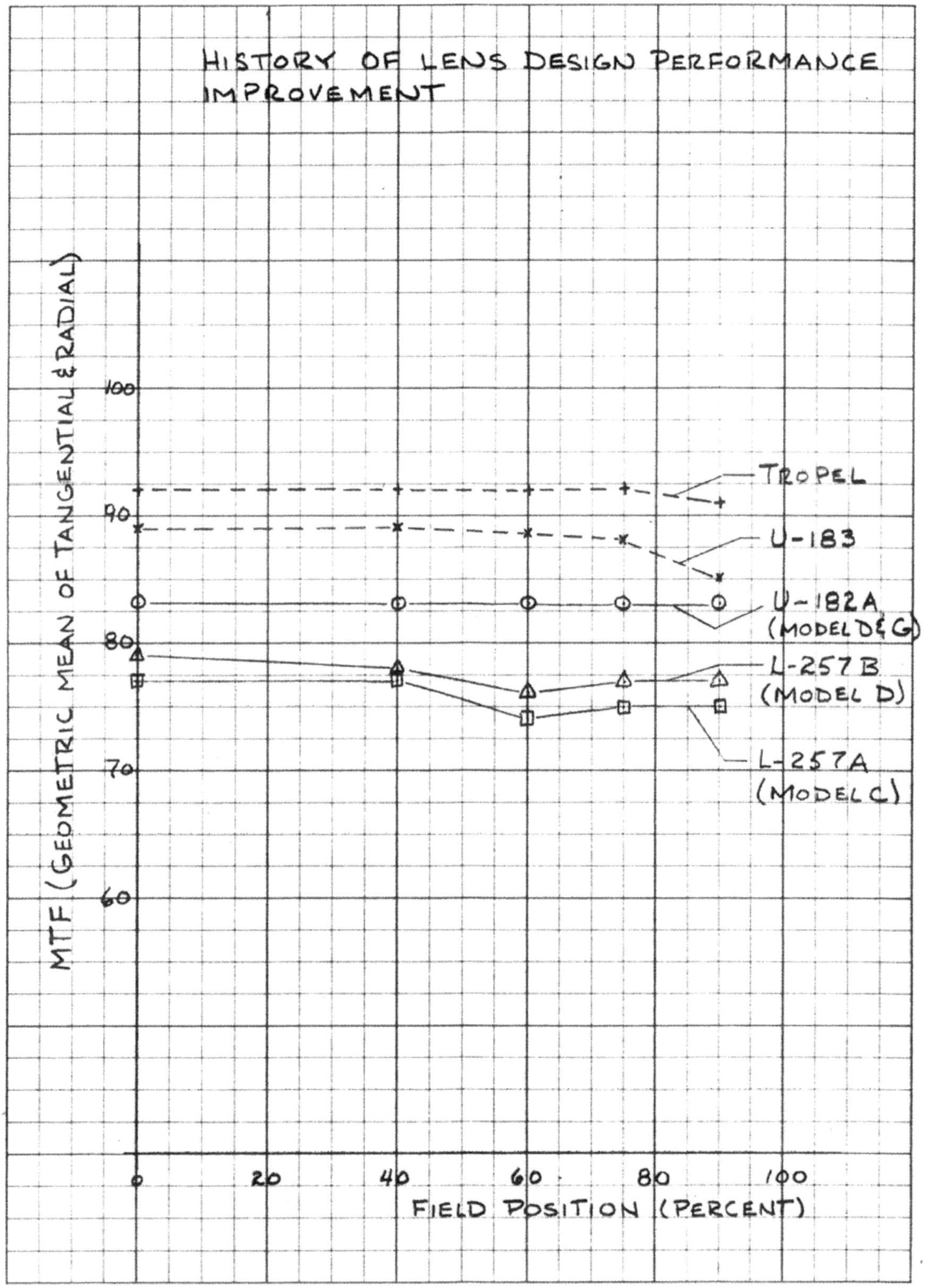

Figure 17: **Lens Design Improvement**

History of lens design improvement. This summary shows the progression of lens design improvement with the Tropel design better than the in-house designs- at least up to that point in time.

1880 Kodak 1980

A 100-year start on tomorrow

June 5, 1981

Mr. J. E. Maher

Subject: Attached P.O. to Tropel

This purchase order authorizes Tropel to fabricate two (2) unit magnification lenses, based on their lens design study completed in May 1980, with agreed to modifications as outlined in the letter proposal from John Buzawa. One of the lenses we agreed to procure for PDD for the P-183 Image Quality Improvement Breadboard (they will reimburse CPD by EKR). The other will serve a dual purpose:

1) as the most advanced apochromatic lens design available, it represents an excellent starting point or baseline for the in-house F-9 design studies to be initiated later this year. Therefore, information gained from an independent assessment and/or demonstration of manufacturing feasibility would have a direct bearing on the approach taken by Optical Engineering.

2) it will be installed on CB-1 for further assessment of optical image quality needs.

The intent of this purchase order was reviewed with W. H. Price and he will inform the Optical Engineering Group.

P. B. Day

PBD:sew

EASTMAN KODAK COMPANY · KODAK PARK DIVISION

***Letter 4:* Tropel Lense Puchase Letter**

Letter, dated June 5 1981, to J. E. Maher requesting authorization to purchase two Tropel lenses. Tropel did in fact fabricate two lenses but it was not a demonstration of manufacturing feasibility. The document with the letterhead celebrating "A 100-year start on tomorrow" certainly is of historic significance reflecting the confidence in Kodak's future.

KODAK APPARATUS DIVISION OPTICAL ENGINEERING DEPARTMENT

FORMULA 56-M-422
EF 494.20
SEMI-FIELD 15.30 DEG.
SEMI-DIAG. 271.75
FORMAT 302MM X 450MM

COPY LENS WITH INCREASED
FORMAT OVER M-420 AND
USING OHARA PSK02
PRINCIPAL WAVELENGTH 500
460-656 ACHROMATISM

F/NO. F/ 5.60
EFF. F/NO. F/11.22
BF 431.21
FF 431.26
1/MAG. 0.997X
OBJ. DIST. -.35
IMAGE DIST. 1.84
BEST FOCUS -1.84

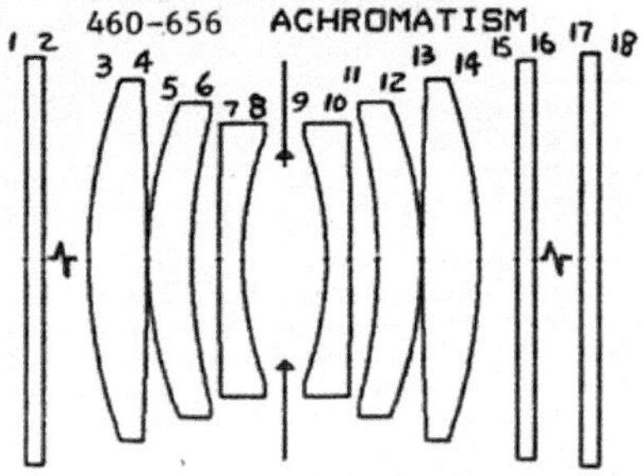

SURF.	CLEAR APER.	AXIAL BEAM	RADIUS	THICKNESS	GLASS	SPECIAL NOTES
1	541.9	0.02	PLANO	5.588	511-604	K-7 (SCHOTT)
2	540.23	0.35	PLANO	919.997		
3	114.60	83.13	182.524	18.986	618-634	PSK02 (OHARA)
4	109.96	81.45	-1387.46	0.504		
5	99.42	79.82	141.494	14.110	618-634	PSK02 (OHARA)
6	90.12	75.38	195.753	9.003		
7	85.06	73.35	2014.67	7.566	613-443	KZFSN4 (SCHOTT)
8	75.83	70.44	119.667	13.762		
	70.44	70.44	DIAPHRAGM	13.762		
9	75.85	70.44	-119.667	7.566	613-443	KZFSN4 (SCHOTT)
10	85.09	73.35	-2014.67	9.003		
11	90.15	75.38	-195.753	14.110	618-634	PSK02 (OHARA)
12	99.45	79.82	-141.494	0.504		
13	110.01	81.45	1387.46	18.959	618-634	PSK02 (OHARA)
14	114.63	83.13	-182.524	12.700		
15	124.69	81.56	PLANO	5.588	511-604	K-7 (SCHOTT)
16	126.35	81.23	PLANO	904.900		
17	541.83	0.67	PLANO	5.588	511-604	K-7 (SCHOTT)
18	543.5	0.00	PLANO			

LENS LENGTH 127.833

PTZ SUM(F/100) = 0.001236

RADII ON SURFACES 3 AND 14 IS A NEW TEST PLATE

IMAGE HEIGHT	ANGULAR FIELD	APPROX. REL. ILLUMINATION	MAX. APER. WITH NO VIGNETTING
93.92	5.40	0.99	F/ 5.6
189.53	10.80	0.95	F/ 5.6
271.75	15.30	0.90	F/ 5.6

PUPIL	LOCATION	DIAMETER
ENT.	62.97	88.23
EXIT	-62.95	88.23

ALL DIMENSIONS IN MM
DESIGNER LRE
ASSIGNMENT 3057-5
SCALE [illegible]

Formula 7: **Lee Estelle Lens Design**

A Lee Estelle lens design in which a moveable meniscus element is added to the Tropel five-element formula, in an effort to provide the same image quality at .767x and .642x magnification. It was not successful. compensated. If the folding mirrors and lens are structurally connected, the entire assembly can be lowered in conjunction with repositioning the lens.

EASTMAN KODAK COMPANY
KODAK APPARATUS DIVISION

FILE COPY DEPT. 89

OPTICAL ENGINEERING DEPARTMENT

FORMULA 77-U-180
OBJ.DIST. 0.0
SEMI-FIELD 18.7° (MAX.)

SIX ELEMENT PANCRATIC LENS
FOR DEDICATED PRINTER

PRINCIPAL WAVELENGTH 546
486.1 - 656.3 ACHROMATISM

OBJ.-IMAGE 660.40
TO CHANGE MAG. ENTIRE LENS IS MOVED WHILE THE ELEMENT COMPRISED OF SURFACES 3 & 4 IS INDEPENDENTLY SHIFTED TO ADJUST THE FOCAL LENGTH.

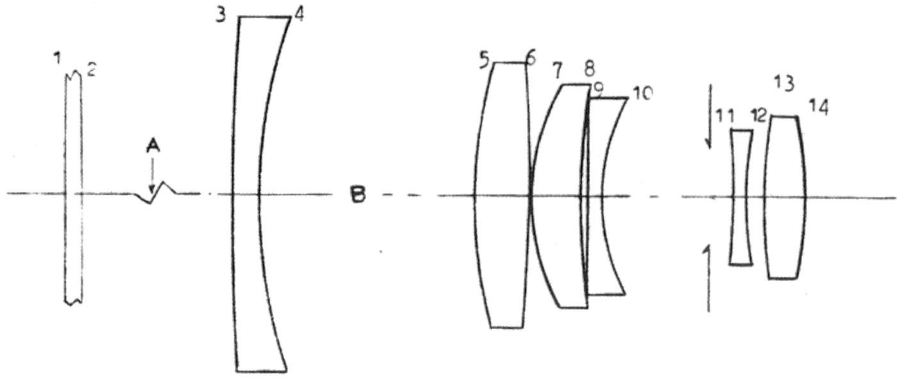

SURFACE	CLEAR APERTURE	AXIAL BEAM	RADIUS	THICKNESS	GLASS	SPECIAL NOTES
1	218.50	14.90	PLANO	3.175	523-586	
2	218.50	15.03	PLANO	A		
3	63.49	15.27	526.127	5.100	517-522	KF-6 (SCHOTT)
4	59.60	15.43	96.3204	B		
5	47.01	20.59	90.5786	10.120	620-603	
6	44.47	20.51	-294.055	.173		
7	39.70	20.42	42.9968	9.000	620-603	
8	35.01	18.64	142.919	1.334		
9	34.64	18.35	PLANO	2.450	550-458	
10	30.16	17.73	38.4175	19.294		
	17.43	17.43	DIAPHRAGM	4.300		
11	20.41	17.37	-132.325	2.500	596-397	
12	22.35	17.49	75.7322	3.354		
13	26.30	18.16	103.666	7.470	620-603	
14	29.07	18.53	-75.7322			

MAG.	E.F.	B.F.	FF(3-14)	F/#	EFF. APERTURE	IMAGE DISTANCE	BEST FOCUS	SEMI-FIELD	IMAGE DIAM.
1.000	160.73	143.22	90.21	8.22	16.44	304.01	-.13	18.68	218.5
.899	160.23	142.97	89.21	8.20	15.58	287.26	-.15	16.96	186.5
1.242	158.60	142.16	85.94	8.13	18.25	339.45	-.30	17.76	228.4
1.667	149.05	137.39	66.80	7.74	20.67	386.89	-.60	18.15	256.2

MAG.	A	B	LENS LENGTH (3-14)
1.000	248.795	39.451	104.546
.899	265.070	39.952	105.047
1.242	211.362	41.612	106.707
1.667	153.769	52.071	117.166

DIMENSIONS IN MM

DESIGNER LRE	SCALE 1X
ASSIGN 6007-1	DATE 12/10/76

KA778 (REV 1-73)

***Formula 8:* Lee Estelle M-Type Formula**

This six element symmetric M-type formula by Lee Estelle became the ultimate lens design solution for the Ektaprint 250 duplicator. However, it was accompanied by a radical change in the way the longer overall object distance is is compensated. If the folding mirrors and lens are structurally connected, the entire assembly can be lowered in conjunction with repositioning the lens.

12

KAARs, SPAARs & Patents

Toward the end of my eleven years in Copy Products I was on assignment to the Research Labs at Kodak Park as a consultant on lenses and optics in general. I remember the rarified atmosphere of having lunch in the cafeteria on the bridge over Lake Avenue connecting the Research Labs with Kodak Park. I considered it a privilege to meet with the researchers who were developing toners, photoconductors and other elements vital to the copy process. It's been too long to remember the names of people who worked directly on the program but just recently I learned of those who did the fundamental research and held patents dating back to the 60's.

This brings me back to the subject of patents and how important they are for constructing the legal basis for doing business. As I look back at the tremendous volume of work I caused or should I say inflicted on Kodak's attorneys I feel deeply chagrined for much of it was worthless, of little practical value. I must have had time on my hands to indulge in these mind games — solving hypothetical problems.

But Kodak made it easy, and in fact, encouraged its employees to submit ideas. These were dignified as advanced reports called KAARs or SPAARs depending on whether they originated in the Kodak Apparatus Division or Special Products. Between April of 1976 and June of 1986 I simply took advantage of the system submitting a total of 34 advanced reports or inventions.

In each case an attorney had to search the patent literature and determine whether to apply for a patent or publish the essence of the idea in the monthly journal, *Research Disclosure*, which went to over one hundred and twenty patent offices and libraries throughout the world. The disclosure simply meant that Kodak had no interest in implementing the idea and no one could claim exclusive rights to the idea if they chose to use it. With a patent Kodak could, of course, demand royalties.

There were 10 Research Disclosures published in my name and a total of 28 patents filed although 20 of those were duplicate filings in various foreign countries. As I said it was a tremendous volume of work for the Patent Department, which I never fully appreciated. Only twice when a patent came through to the boss, were my contributions recognized with a letter of commendation. Not that it mattered to me then, but earlier in my career I had submitted a patent application and never heard back one way or the other if the company acted on it. This was the North American TV sight I had worked on just prior to my transfer to Research and Engineering.

Fast-forward twenty-two years into my retirement, and for nothing better to do, I launch into writing my memoirs. In the back of my mind I knew that I had patents, there was some evidence, letters and correspondence with attorneys but no patents. It occurred to me that there might be someone in the Patent Department at Kodak, that company I once worked for, who might be willing to search out my patents. So I called and got Beth Read, Manager, Patent Operations, Office of IP Counsel. She was most helpful and within a day had sent all seven of the patents in the Kodak database. I was amazed but still no sign of a patent for the TV bombsight. The next day she was able to access the Navy patents on the internet and there it was. This was the one innovation that was successfully implemented on an important contract with the Navy.

Just to complete the story of my inventions actually implemented, there was the one that a Japanese firm, Ricoh, picked up from the patent literature and incorporated in a desktop copier. Did Kodak obtain royalties? Probably not. A Ricoh engineer claimed credit for developing the roof mirror lens array, no mention of my patent, in a July, 1985, article in the *Technology Journal.*

It was fascinating to note that they had made MTF measurements at 5 lines/mm of 70 to 80 percent with a depth of focus plus or minus 0.6 mm — imaging characteristics similar to those I had specified for the Kodak Ektaprint copiers.

"The Nation thinks small and aims low"

This was the headline of a 2007 article on the editorial page of the Democrat and Chronicle by James P. Pinkerton. "The world is at least as dangerous in 2007 [and today 2012] as it was in1957."

> "Fifty years ago this week, the Soviet Union stunned the world by orbiting a space satellite, Sputnik.
>
> In 1957, Americans felt that we might lose the Cold War if the Russians could gain the "high-frontier" of outer space. Fortunately, our political system rose to the challenge, providing the world with a lesson in the power of good leadership and mobilization in a free society.
>
> Back then, the 34th president was determined to gather the tools to do the job right. Dwight Eisenhower knew that we needed more than rockets to out-fly the Soviets; we also needed scientists who could out-think Moscow. Within a year of Sputnik, Ike signed the National Aeronautics and Space Administration into law; in addition, he inked the National Defense Education Act, establishing science training as an urgent priority.
>
> And it worked — barely more than a decade later, in 1969, America put two astronauts on the moon [and incredibly sophisticated reconnaissance satellites orbiting the earth]. Uncle Sam had won the space race."

JuneMon01 Apr06 07, 0520082006

From: beth.read@kodak.com
To: <pierceandsally@rochester.rr.com>
Date: JuneMonday01 April06 07, 0520082006 08:38 AM
Subject: US Patents

Mr. Day,

I've located the following US patents in our database that issued with your name on as sole or contributing inventor:
US 4,338,021
US 4,427,291
US 4,371,253
US 4,428,662
US 4,459,011
US 4,477,177
US 4,740,818

Attached are copies of each of these patents:

In addition, I am attaching a report showing the corresponding foreign patents that also issued:

It will take me a little bit longer to try to search for the patent that issued in 1958. I will get back to you.

Beth Read
Manager, Patent Operations
Office of Chief IP Counsel

Eastman Kodak Company
343 State Street
Rochester, NY 14650-2201

beth.read@kodak.com
Office: 585 722-9893
Fax: 585-588-7413
us-patent@kodak.com

Kodak

Page 1 of 1

United States Patent [19]

Day

[11] **4,427,291**

[45] **Jan. 24, 1984**

[54] **COMPACT OPTICAL IMAGING APPARATUS**

[75] Inventor: **Pierce B. Day**, Pittsford, N.Y.

[73] Assignee: **Eastman Kodak Company**, Rochester, N.Y.

[21] Appl. No.: **376,472**

[22] Filed: **May 10, 1982**

[51] **Int. Cl.**[3] **G03B 27/44**

[52] **U.S. Cl.** **355/46**; 355/1; 355/51

[58] **Field of Search** 355/46, 51, 50, 1

[56] **References Cited**

U.S. PATENT DOCUMENTS

3,606,535	9/1971	Koizumi	355/51
3,655,284	4/1972	Agliata	355/46
3,836,249	9/1974	Weber	355/51
3,982,833	9/1976	Kolibas	355/51
4,316,665	2/1982	Mochizuki et al.	355/46
4,357,099	11/1982	Taguchi et al.	355/46

Primary Examiner—Monroe H. Hayes
Attorney, Agent, or Firm—Lawrence P. Kessler

[57] **ABSTRACT**

Compact optical imaging apparatus for projecting in focus an upright and wrong-reading light image of an information bearing document in an object plane onto a parallel image plane. The apparatus comprises a plurality of lenses located between the object plane and the image plane for projecting a plurality of overlapping segments of a light image along an optical path. A light image of an information bearing document in the object plane is directed to the lenses. A plurality of roof mirrors (i) receive such image segments from the lenses, (ii) reverse and invert such image segments, and (iii) reflect the image segments back through the lenses along the path toward the image plane. The image segments are then directed from the lenses to the image plane to form an upright and wrong-reading image of the information bearing document in the image plane.

6 Claims, 5 Drawing Figures

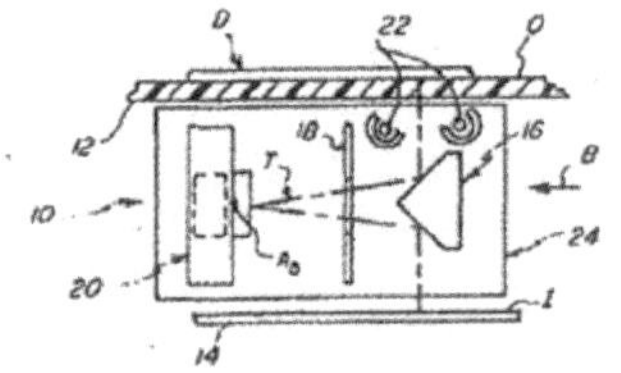

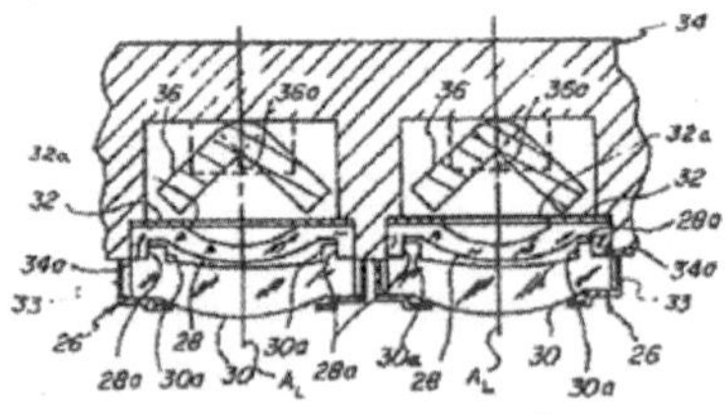

U.S. Patent Jan. 24, 1984 Sheet 1 of 3 4,427,291

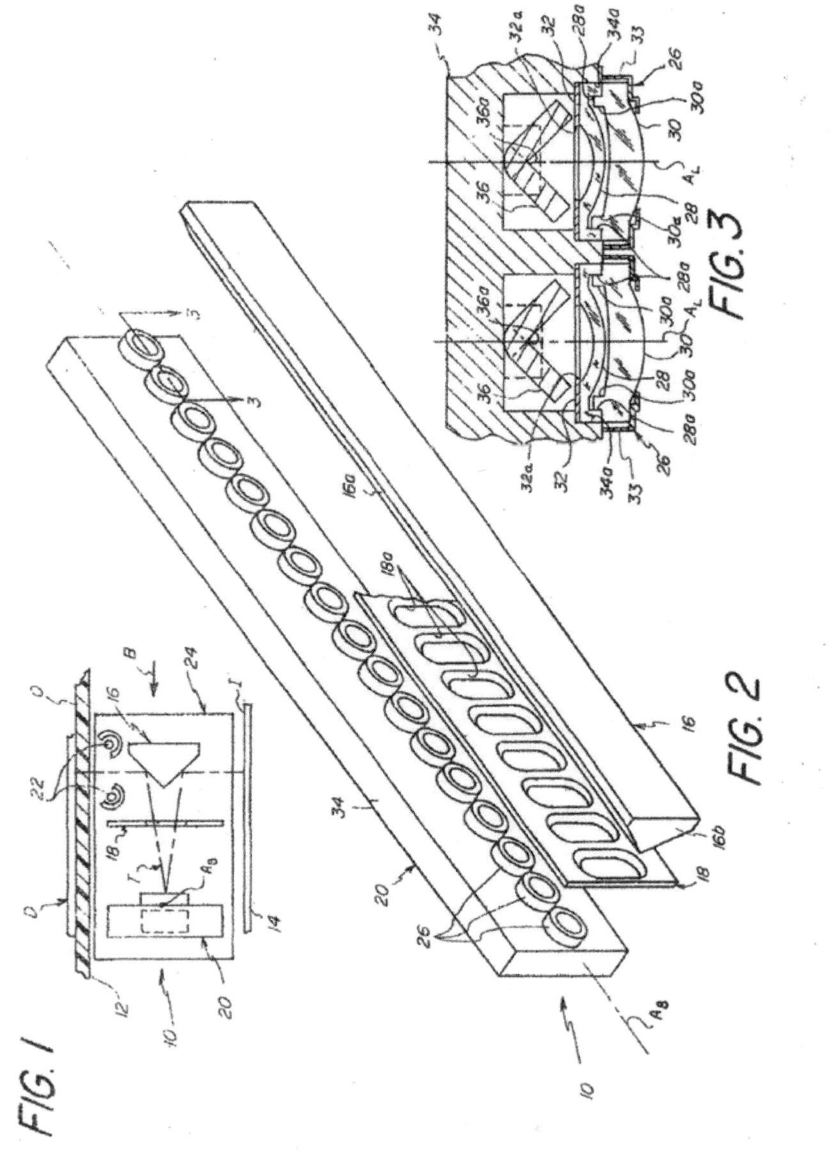

Technology Roof mirror lens array modules

Imaging Properties of Roof Mirror Lens Array Modules In Office Equipment Application

By Motoaki Kawazu,
Ricoh Company Ltd.

As unit magnification imaging devices, a gradient index rod lens array and a strip lens are widely used in copiers[1][2]. These imaging devices form synthesized images continuously by a micro lens array. Compared with conventional copying lens, there is an advantage that object-to-image distance can be greatly reduced on effective scanning width also.

In office equipment, copiers, facsimiles and non-impact printers and the like are smaller, for personal use. This increases the demand for imaging devices.

The RMLA, or roof mirror lens array introduced by the Ricoh Company at the COMDEX In Japan '85, held in April in Tokyo, was developed for such purposes.[3]

Principle and Structure[4]

The RMLA consists of a lens array (LA) and a roof mirror array (RMA), formed into one, long, plastic molding. The RMLA forms a normal image directly, without forming an intermediate image.

The RMLA has an advantage in that the recurrent reflection function and the summetrical lens system provide stable imaging properties against dimension fluctuation that result from temperature and humidity changes. Overlapped images are continuously formed in an array direction to get a whole image.

Figure 1 shows the relationship between a single roof mirror lens and an object-to-image. Figure 2 shows the basic structure of the RMLA. A whole, normal image is obtained by overlapping images in the Y direction continuously, and an inverted real image is formed in the X direction.

An object-side optical path is separated from an image-side optical path by a placing a pair of optical separation mirror (SM) in front of LA. Figure 3 shows the overall structure. Conjugate length (L) at a required value can be selected by optionally setting the space between two pieces of optical separation mirrors and LA front.

Figure 4 shows the overall structure in X direction cross section. Numbers and positions of SM which are in front of LA, can be changed in accordance with product requirements.

Since each lens of LA and a roof mirror of RMA are continously arranged at a constant interval, a single, plastic block molding becomes possible.

Design Values[5]

Figure 5 shows an equivalent structure diagram of a single roof mirror lens optical system. In

Fig. 1: Image formation of the single roof mirror lens

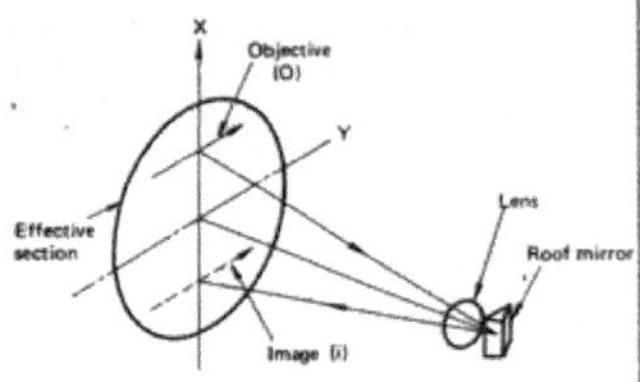

Fig. 3: Overall structure

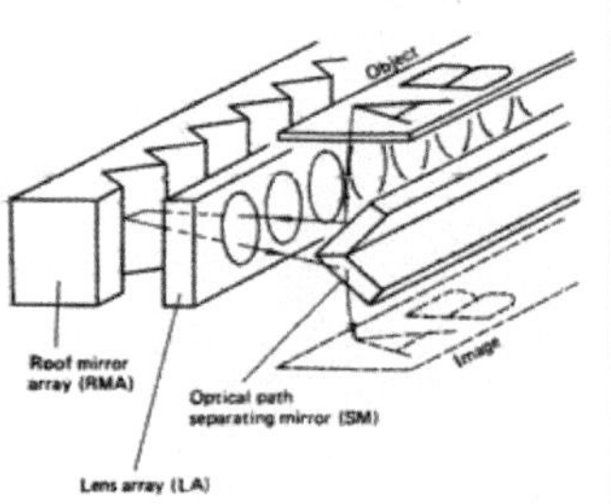

RMLA, combining unit magnification imaging functions by symmetrical lens and recurrent reflection functions of a roof mirror, a normal real unit magnification image in Y direction is formed. In setting these optical purposes, it is necessary to secure performance of the lens system and to rationalize MTF (modulation transfer function — well-balanced aberrations) of the composed lens system and irradiency distribution. It is necessary to consider the restricted conditions from the point of the molding process.

Since RMLA forms unit magnification images in the symmetrical lens system (equivalent four faces) without forming an intermediate image, high-irradience and high-resolution can be established in the design.

Table 1 shows examples of theoretical designs.

Characteristics of RMLA[6]

Characteristics are measured in accordance

Fig. 2: The basic structure of an optical system and an image formation in RMLA

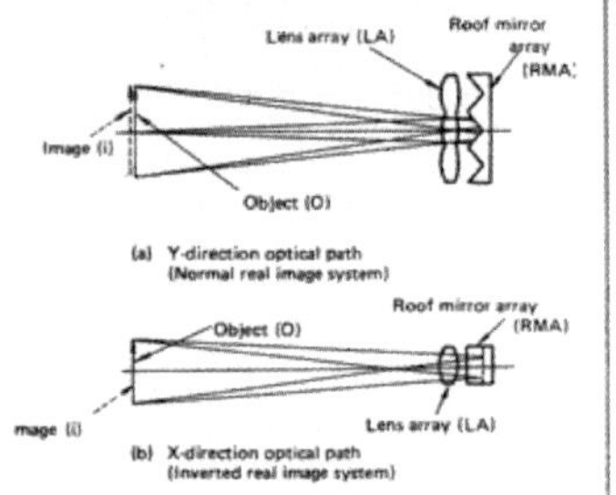

Fig. 4: Structure of X-direction

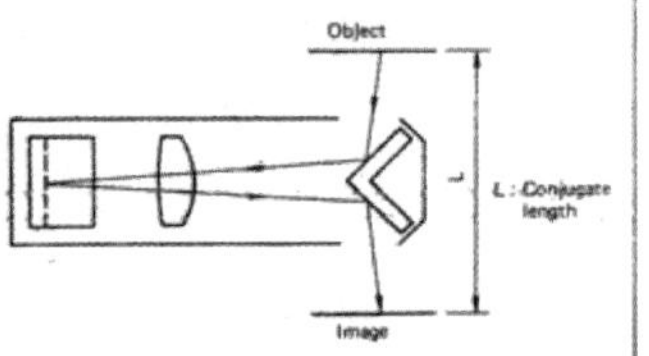

with the examples of design in the table.

MTF

Figure 6 shows results of MTF measured with Ricoh's automatic measuring system. Due to the difference of overlapping by places, periodic changes of MTF can be observed. At MTF of 5 lp/mm, approximately 80 percent in X direction and approximately 70 percent in Y direction are available.

Depth of Focus

Figure 7 shows changes of average MTF in 5 lp/mm when images are defocused. The territory over 40 percent of MTF is ±0.6 mm or more. Being a single array, compared with Y direction, the depth of focus in X direction shows a much larger value.

Figure, 8 shows changes of average MTF when objective surface and image surface are equally defocused. In this case, the apparent depth is great. Figure 9 shows projected image patterns.

Appendix A

Large Optic Metrology Development at Kodak

By John Shafer[95]

(Edited and condensed with the author's permission)

Introduction

The United States Government recently (September 2011) declassified several "Spy-in-the-Sky" satellite programs and Eastman Kodak Company's involvement in those programs. This document will present a history of large optic metrology development that covers over ten years of optics manufacturing development (for Gambit and Gambit 3 programs), within the Eastman Kodak Research and Engineering Organization.

The author of this document, John Shafer, was involved in these programs for a period of approximately ten years, with most of his effort involved in optical testing, optical metrology development and manufacturing process development. It was extremely interesting to the author because of the many hours each week and the years spent in technology development required to fabricate the Gambit and Gambit-3 mirrors.

The following outline provides a listing of various areas covered in the approximate chronological order in which work progressed starting at the beginning of the Gambit Program.

The Gambit Program

- Test glass technology
    - Radius control
    - Power and irregularity
    - Master Test Glasses
    - Working Test Glasses
- Foucault Knife Edge Testing
    - Topographical
    - Very Slope Sensitive
    - Subjective
- Auto-collimating Point Source Microscope
    - Subjective
    - Analytical for diffraction limited images
    - Quantitative for astigmatism
- Interferometry
    - Laser unequal path development
    - Interferometers with two optical systems in one
    - Evaluation of interferograms
    - Calibration
- Spherical Mirrors
- Flat Mirrors
    - Zero Power Tolerance
    - Ritchey Test Setup with Master Sphere
    - Correlating two orientation tests
- 300 inch Collimator
    - Tri-bar Resolution Testing

[95]John Shafer. B.S., Ohio State University, 1962; M.S., University of Arizona, Optics Science Center

Quantitative "Go-No-Go" test
Not useful in analyzing failures
Difficult to calibrate readers
Meniscus Matching

GAMBIT-3

The Primary Mirror
32-foot vertical test facility
Refractive Null Corrector
Defining Optical Axis
McLeod Axicon Telescope
Airbag support
Isolating test set-up from vibrations
The Stereo Mirror
Ritchey Test Setup
Interpreting the Interferogram
Complete Lens Assembly Evaluation

GAMBIT-PROGRAM OPTICS DEVELOPMENT

A Brief Summary

The Gambit lens was a 77-inch f/4 Maksutov consisting of a 28 inch diameter spherical mirror, a 20 inch diameter meniscus lens located at the stop of the system and a package of field flattening elements near the focal plane. A 33 inch diameter flat mirror was positioned in front of the lens for forward and aft stereo imaging. Initially the complete system was to be built to a tri-bar resolution specification of 100 lines/mm which was equivalent to 3 foot ground resolution, when taken from 95 miles in orbit.[96]

The author's first assignment on this program was that of Quality Control engineer in 1962 evaluating the final product delivered by the Gammbit Lens Assembly Department.[97] The three optics received from that department were a 28 inch diameter, light weight (egg-crate construction) spherical mirror with a 168 inch radius of curvature, a 33 inch diameter light weight (egg-crate construction) flat mirror and a 20 inch diameter solid meniscus lens.

In the early days the only tool used for evaluating mirror quality was an Auto-collimating Point Source Microscope. When testing the 28-inch diameter spherical mirror the instrument was focused at the center of curvature. An acceptable spherical mirror would produce a reflected "star" image with a complete diffraction ring pattern.

A Ritchey Test was used to evaluate the flat mirrors. In the test setup light from the Auto-collimating Point Source Microscope was reflected off the flat being tested at 45 degrees to a master quality 28-inch diameter sphere. After a double bounce off the flat, the reflected "star" image should have a perfect diffraction ring pattern. If there were any power in the surface of the flat it would show up as an astigmatic image. The distance between the horizontal line image and the vertical line image could be used to quantify the magnitude of the power present in the flat.

[96]Readings in the range 130 to150 lines/mm may have been recorded, but no records survived.
[97]All optics manufacturing and assembly took place at Hawk-Eye.

After acceptance of the two lightweight mirrors by the Final Product Assembly Department, they were assembled into lightweight metal mounts using a low-shrink adhesive and retested. To minimize distortion caused by shrinkage of the adhesive, it was applied in segments around the periphery of the mirror.

After the Final Product Assembly Department completed the assembly of the lens and folding mirror, the optics module was tested with the Auto-collimating Point Source Microscope by auto-collimating off of a master flat. If the image quality was determined subjectively to be acceptable, the lens and optics modules were submitted for final evaluation on one of two 300-inch Focal Length Collimators. Test cameras were used to record the images of the standard 1951 Air Force Tri-bar Target located on-axis and various field positions.

Through focus tests were used to determine the plane of best focus for mounting the camera. If the surface of best focus was curved, adjustments could be made with the field flattening optics.

Large Optic Component Fabrication

Primary Mirror

The primary mirror was a 28-inch diameter lightweight spherical mirror with the 168-inch radius of curvature controlled to about one thousandth of an inch. However, the surface figure tolerance of 0.1 wave peak-to-valley (2.5 millionths of an inch) was extremely difficult to meet considering the tools available, the size of the mirror and its egg-crate construction. The egg-crate mirror blanks, developed at Corning, had thin front and back fused silica plates that were adhered to an internal crosshatched lattice structure.

Well before the start of the polishing operation all the test glasses for the program had to be ground and polished. The first of these was the 8-inch diameter solid master test glass test. Kodak's Optical Manufacturing Organization had many years experience in manufacturing test glasses for other programs. It maintained a master test glass vault, which was kept under lock and key that contained thousands of master test glasses of various sizes and radii, both concave and convex. However, test glasses had seldom been made of such a long radius of curvature and of such a large diameter.

After the concave master test glass was made and certified to the required radius, multiple convex working test glasses had to be made for daily use in production. It was soon found that test glasses of this diameter had to be beveled because when in "optical contact" the test glass was very difficult to remove. Perhaps the first major challenge for the Large Optic Fabrication Department was to provide to the opticians these working test glasses to the same exact radius and to a surface tolerance equal or better than 0.1 wave.

The first phase in polishing the 28-inch diameter primary mirror involved figuring and testing the mirror on the polishing machine. In past experience with test glasses roughly equal to the part being polished, a surface fit of 1/2 wave irregularity (1 ring) was considered very good quality. However, the task of polishing the 28 inch diameter primary mirror to 0.1 wave irregularity with an 8 inch test glass presented a whole new order of difficulty. Since the entire surface could not be viewed at the same time, the only viable approach was to assume annular symmetry and use the test glass to "profile" the surface figure near the edge of the mirror.

As the mirror quality approached the 1/4 wave fit (1/2 ring) it would have to be removed after each polishing operation and transferred to a test area for a more effective evaluation. Because the polishing operation of moving a polishing lap across the surface put heat into the mirror, it was necessary to let this heat dissipate before each test.

Foucault Knife Edge Testing

The 28-inch lightweight mirror under test was set up on edge, in a wood v-block, on a vibration-isolation table with the Foucault Knife Edge Test instrument[98] placed 168 inches from the mirror at its center of curvature. In the usual configuration a pinhole light source and small folding mirror is offset to one side of the primary mirror centerline and the knife-edge is offset to the other side at the center-of- curvature. If the operator views the reflected light from the mirror by placing his eye just behind the knife-edge, he would see a topographic shadow-image of the mirror surface. Very small deviations, fractions of a wavelength, from a perfect sphere could be observed as light and dark shadow areas.

However, the knife-edge test map of the surface was very difficult to interpret because the shadow pattern varied depending on the longitudinal position of the knife-edge. For example, if the mirror had a slightly longer radius of curvature for the central portion of the mirror than that of the edge annulus, then a knife-edge positioned at the center of curvature of the central portion would show a perfect central area and a high edge. Moving the knife-edge in toward the mirror, the outer annulus would look perfect, but the center would look high. Depending on where the optician wanted to remove material, he had to pick a point of reference for the knife-edge test. For this reason a Zonal Test was developed to make sure the optician always used the center area as the starting point.

The iterative process of polishing and testing mirrors was far from an exact science and depended on the skill and experience of the optician. It could take many weeks to produce an acceptable quality mirror and often under extreme schedule pressure the assigned optician had to be replaced by a more senior optician.

The Ritchey Test

Testing the 33=inch diameter stereo mirror called for a Ritchey Test set-up with a master sphere. The mirror was positioned on its edge in a v-block on a vibration-isolation table angled at 45 degrees to the centerline of the master sphere.

As in testing the 28-inch primary mirror, surface imperfections were evaluated using the Foucault Knife Edge test method. But the interpretation of the shadow pattern was far more complicated.

In the first place any power in the surface would produce an astigmatic image. Opticians had been working to a "zero power" 0.1 wave irregularity tolerance- an almost impossible requirement.

[98]Commonly used by amateur telescope makers for figuring small astronomical mirrors.

To further complicate matters, this one-orientation test had an inherent built-in error because of the apparent shortening of the horizontal diameter of the mirror viewed at 45 degrees. Any component of power and/or cylindrical error in the surface produces twice the wave-front error in the vertical direction as in the horizontal direction but would only be observed as a focal shift. Since the astigmatism (or lack thereof) could not be easily observed in the knife-edge test, it was necessary to switch to the Auto-collimating Point Source Microscope to assure that an astigmatic image was not being produced. The Foucault Knife Edge test could then be used to attack the remaining surface imperfections by combining the two orientation test results. If this wasn't complicated enough, the master sphere was only 28 inches in diameter and only a portion of the 33-inch flat could be viewed at the same time.

This was the state of optics technology when these lightweight egg-crate mirrors were manufactured for the Gambit program. Interferometric testing was in development in 1965 but was not yet available for testing mirrors.

However, the first application of interferometry was on the first Gambit lens assembly. It may have been at that point that when viewing the auto-collimated wave front, another problem was diagnosed. The wave front exhibited a radially symmetric turned down edge, which was traced to an index variation in the 20-inch meniscus lens. In production meniscus lenses could be paired with a primary mirror and shipped as a matched pair. However this was not always successful and it was necessary to resort to aspherizing the individual meniscus lenses.

Organization and Politics

Before the start of the Gambit program all optics manufacturing was done within the Consumer Optics Manufacturing Organization including the optics for Samos. When the Large Optics Fabrication Department was formed, all the opticians were recruited from consumer optics where there was this mystique about the art of polishing optics. They didn't want help from anyone looking over their shoulders. Their decisions were not easily questioned.

As director of Kodak's Apparatus Research and Development Organization as well as the Research and Engineering Division at Lincoln Plant, A.B. Simmons was responsible for the design and management of the Gambit program. He was concerned about the non-scientific tail of the Large Optics Fabrication Department wagging the dog of a more scientific Research and Engineering Organization. He created a new organization called Optical Operations to help develop optical technology throughout all organizations. It was staffed with a variety of talent to address specific optical problem areas.

John Shafer was assigned as the development engineer to help the Large Optics Fabrication Department develop better optical metrology capabilities. He spent considerable time working with the opticians becoming familiar with their practices in order to gain their confidence. Among his many innovative contributions to the program were: a Zonal Test Procedure for use with Foucault Knife Edge testing; development of a fixture and procedure for meniscus matching; and initiated a procedure in which the meniscus lenses were aspherized by hand-polishing.

Interferometry

In 1965 after several years experience on the Gambit program, John Shafer was called into Art Simmons' office and given the assignment of determining how interferometry should be applied in support of the Research and Engineering Organization. It was not long after the development of the HeNe gas laser [99] that it found application as the source for the unequal-path Twymann-Green interferometers. The coherence of laser light made it possible for the length of the test arm and the reference arm to be vastly different. For example, in the reference beam a plano wave front is bounced off a small master flat; in the other beam a microscope objective focused light to a point from which a perfectly spherical wave front is produced so that a spherical mirror of any size could be tested. When the return wave front carrying any surface irregularities is combined with the reference wave front, high contrast fringes were produced over the entire surface of the optic under test. The photographic record or interferogram could then be scanned into a computer. This development and application of the Twymann-Green Interferometer to large optic fabrication would be crucial to the future success of the Gambit-3 program.

John Shafer was again called into Simmon's office and asked to write tthe charter for a new organization (the Interferometric Test and Evaluation Group) for the development and use of laser interferometry. There were three sections: a continuing metrology development section; a software development section for automating the interferogram evaluation process; and a production section to conduct all interferometer production tests and evaluations.

Most early interferometer evaluations were done by hand. One of the first challenges was to produce acceptable quality photographic images, which could be scanned automatically for computer evaluation. The problem with the commercially available interferometers was the fact that they didn't produce images with sharp edges, of the optic being tested and surface fiducial marks were poorly defined. After considerable frustration the metrology development section came up with a fix, a modified interferometer which added relay optics to collect light back-scattered off the optic under test. An image of this scattered light is formed between the objective lens and the beam-splitter. Relay optics transferred this scattered-light image to the film-plane of a 35mm Retina camera. Interferograms then showed sharp, high contrast fringes right out to the edge of the optic and sharp fiducial marks.

As interferometers were being placed in the optics fabrication and assembly areas, it was important for the opticians to learn set-up procedures and how to visually evaluate fringes. John Shafer wrote the training manual and directed a series of courses available to program peronnel.

To the opticians the interferometer was essentially an infinitely variable radius, non-contact test glass, and it was quickly accepted for the significant benefit it provided.

[99]Theodore Maiman is credited with the invention of the ruby laser in 1960. The HeNe gas laser, invented by Ali Javan, became available commercially in 1961.

GAMBIT-3 OPTICS DEVELOPMENT

At the start of the Gambit-3 program in January 1964, the Large Optics Fabrication Department was already committed to a delivery schedule on Gambit-1 of one-a-month plus the Lunar Orbiter for NASA. Completely separate facilities were needed for a greatly expanded optics fabrication and assembly operation. The Interferometric Test and Evaluation Group was responsible for developing the new metrology and mirror handling procedures required. Continuing development of interferometric testing was absolutely crucial in meeting end product specifications.

The Primary Mirror

The primary mirror was a 44 inch diameter lightweight asphere with a 32 foot vertex radius of curvature. Preliminary testing of the mirror before aspherizing showed that mounting the mirror on its edge on a v-block with the optical axis horizontal was not feasible. Because of its size the mirror sagged under gravity causing a surface distortion beyond the required .05 RMS wavelength tolerance. In addition, horizontal testing of the mirror, with a 32-foot radius of curvature, was significantly affected by building vibrations and wave front distortions duc to thermal stratification in the long air path.

The first problem for the Large Optic Fabrication Department was where to locate a 32-foot vertical test facility. A stiff mechanical structure was required to accurately position the mirror under test and the test instrumentation. The entire structure would then have to be suspended by damped springs to isolate it from building vibrations.

There were essentially two options. One was to build a separate stand-alone enclosure in the courtyard adjacent to the main entrance to Hawk Eye. Building managers strongly objected. In considering the other option they were overruled. It was found that by abandoning an existing elevator shaft right next to the fabrication department, the test facility could be set up inside with access on the first floor and the third.

A significant design effort went into the problem of positioning and supporting the asphere for testing. Various multi-point support designs were evaluated but there were concerns that rigid points of contact would show through and distort the front surface. The effort then turned to an air bag support system, which allowed the mirror to be strain free. At the same time it provided a lifting mechanism and precision alignment to the spring-suspended mechanical structure. Before raising the mirror into contact with the three mechanical-stop-datum on the structure, the air bag had to be inflated to the point at which the rear surface of the mirror was tangent to the air bag. Procedures had to be developed so that this process was repeatable.

The Refractive Null Corrector

The interferometric testing of the aspheric primary mirror could best be done using a specially designed Refractive Null Corrector placed between the mirror and the interferometer. The design consisted of two elements about ten inches in diameter. When placed at the proper distance from the center of curvature of the mirror, the Null Corrector makes the asphere look like a sphere. The optician could then see surface errors as if polishing a sphere. In fact before aspherizing the mirror, it had to be polished without the null to produce a sphere with the 32 foot radius of curvature.

The Axicon Telescope

As critical as the longitudinal positioning of the interferometer and the Null Corrector relative to the mirror was, the lateral positioning and alignment to the aspheric axis of the mirror were equally critical, if not more difficult, to establish. The first problem, which was unique to polishing aspherics, was how to define the aspheric axis and then align the Null Corrector to it. In Kodak's optical tool-box opticians had an instrument called an Axicon Telescope. The heart of the instrument was a polished flat cone 2 to 3 inches in diameter. A small source of light on the axis of the cone is projected as a bright line in space. It also focuses any light along the axis to the same image point. The reflections off surfaces of a lens, for example, appear as bright dots, which can be aligned with lateral adjustments of the lens elements.

At the start of mirror fabrication a small hole or indentation was drilled in the physical center of the mirror as it rotated on the polishing table. The second point, which was fairly easy to locate, was the center of curvature of the mirror prior to aspherizing. By placing small "grain of wheat lights" (less than 1 mm in diameter) at these two points, they defined the aspheric axis when viewed through the Axicon Telescope. A third "grain of wheat light" was placed on the aspheric axis midway between the mirror and the center of curvature. Then the Axicon could be placed between the mirror and the third "grain -of-wheat light" looking back at the other two lights.

Next, the Null Corrector was brought into place at the proper distance from the aspheric mirror using an invar spacer rod and then aligned to the aspheric axis using the Axicon Telescope. Multiple points of light reflected off the Null Corrector lens surfaces all had to be perfectly aligned to the aspheric axis previously established. Finally the inerferometer was set up on the aspheric axis moved to the proper distance from the Null Corrector, again using an invar rod.[100]

In order to test the mirror, it had to be moved physically from the polishing machine to the air bag support then transported to the test tower. Positioning the mirror up against the three mechanical-stop-datum guaranteed alignment time after time for each test setup. Because of the size and weight of the mirror, it could not be moved by hand. Pressray Air Rings were designed to be lowered around the mirror by a crane. Once in position around the circumference of the mirror, an air tube was inflated which squeezed metal bars against the circumference of the mirror with sufficient force to allow it to be lifted by the crane.

To lift a 44 inch "egg-crate" light-weight mirror without crushing it was just one of many challenges faced by the Ground Support Design Organization. They were charged with the daunting task of procurement, design and construction of all the handling equipment required for the production of the Gambit-3 lenses.

[100] It was this step in polishing the Hubble mirror that became the focus of the committee investigating the cause of the fuzzy images returned from space.

The Stereo Mirror

The polishing and testing of the 58 by 46 inch elliptical lightweight stereo mirror presented another set of complex engineering problems for the Ground Support Design group. First the requirement for handling the significantly larger elliptical mirror called for Pressray Air Ring equipment and cranes in both the polishing and testing areas. The design and construction of the Ritchey Test Setup was a formidable task involving a large mechanical structure suspended on damped-springs to isolate it from building vibrations. It had to support the stereo mirror under test, the interferometer and a massive 71-inch diameter master quality solid sphere suspended at 45 degrees.

Because there were two lens designs for Gambit-3, there were two elliptical stereo mirror configurations. For the earlier units the mirrors had a large 45-degree bevel leaving one end of the mirror structurally weak. Later units, incorporating the R-5 lens design, called for a large rectangular hole in the center of the mirror with a weak tapered section along one edge. In both cases these weak sections had to be supported with glass inserts or plugs during polishing and testing operations and then removed.

For interferometric testing, the mirrors were supported on their backs on air bags eliminating surface distortions inherent in the on-edge testing of the Gambit stereo mirror. This meant that the plane containing the optical axis of the master sphere was vertical rather than horizontal and the operator had to stand on a high platform looking down through the interferometer. A separate area in the basement of Hawk Eye was required with enough space to accommodate an on-edge test of the 71 inch master sphere. (The interferometer was focused at the center of curvature 32 feet from the mirror). Residual surface errors in the master sphere were later backed out, but still there was a serious problem of supporting the mirror at 45 degrees without introducing some surface distortion.

Consultants from outside Research and Engineering, are credited with developing the design of a truly unique strain-free mounting system. The supporting structure at the back of the mirror, in the form of three trusses, was attached to the mirror through pads glued to the rear surface so that the mirror is "virtually supported at three points all in the plane through the center of gravity of the mirror.[101]" Incorporating this mounting system into the design and completing the construction of the Ritchey Test Set-up were major milestones in the program.

[101] J.H.McLeod and M.J. Harrison, "Mounts for Optical Mirrors", December 19, 1962

Interpreting the Interferogram

The Ritchey tests on the Gambit stereo mirror were done visually, as described earlier, and had to account for the apparent foreshortening of the mirror viewed at the 45-degree angle of incidence. Ritchey tests on the Gammbit-3 stereo mirror presented a similar but significantly more complex set of problems for the Interferometric Test and Evaluation Group to address.

As in the Gambit test, the Gambit-3 stereo mirror was rotated 90 degrees about a vertical axis and the test repeated. Fiducial marks were required both on the stereo mirror and the master sphere in order to register the three interferograms (two stereo mirror tests and one master sphere test) and produce an accurate map of the mirror surface for the opticians in the Large Optics Fabrication Department.

There were two problems unique to the Ritchey Test geometry. First, each fringe deviation from a straight line did not represent 1/2 wavelength surface error, as in the straight on testing of a spherical mirror, but 1/2.828 wavelength. Several discussions were required to convince everyone that each bounce of the wave front off of the mirror induced an error of 1.414 times the mirror surface error and hence 2.828 wavelengths of wave front error represented 1 wavelength of mirror surface error.

The second problem was how to combine the two interferometric tests when in one test the 58 inch dimension of the mirror is foreshortened and in the other test the 46-inch dimension is foreshortened. Optical and software experts from across Research and Engineering were pulled together for a two-week Computer Development Task Force.[102] Considerable additional effort went into software development with the end result a rather complex procedure for producing a topographic map. The Software Development Group then converted this activity into a production worthy computer program.

Later in the program, the Interferometric Test and Evaluation Group developed a method for calibrating the interferometers so that even this small but significant wave front error could be backed out.

Clearly the polishing of both stereo and primary mirrors to a 0.05 RMS wave tolerance would not have been possible without the evolutionary development of interferometric testing and the power of the computer to combine, subtract and statistically average interferograms.

Complete Lens Assembly Evaluation

The optical quality requirement for the complete Gambit-3 Lens Assembly was written in terms of a single-pass wave front error no greater than 0.10 wave RMS or an Optical Quality Factor of 70 percent. This was because the double-pass test results could easily be converted to a single-pass measure of performance, which was not true of a tri-bar test.

Once the primary and stereo mirrors were accepted by the Quality Control Department, they were delivered to the Gambit-3 assembly areas first at Hawk-Eye and then at Lincoln Plant. At some point prior to the R-5 lens upgrade, final-assembly testing took place at Hawk-Eye.

[102]Led by Dr. Edward M. Granger, B.S., 1957; M.S., 1969; Ph.D., 1975, University of Rochester's Institute of Optics.

Three large vacuum chambers on isolation pads were installed with instrumentation for interferometric testing in vacuum. These tests on a complete optics assembly were by auto-collimation off of a 48-inch master flat which could be rolled into the chamber along with the unit being tested. Prior to evacuating the chamber, an instrument, known only at Kodak as an "Extensometer", was lowered through a hole in the top of the chamber, like an inverted periscope, with an interchangeable eye-piece and camera at the top end and interferometer optics (microscope objective, beam-splitter and reference flat) built-in at the bottom. The interferometer could be focused at the axial focal point of the lens, projecting a spherical wave front out through the Ross corrector optics to the primary mirror, the stereo mirror and the master auto-collimating flat. After the second pass through the optics assembly, the wave front is matched against a perfect reference wave front forming a double-pass interference fringe pattern. The "Extensometer"optics served to relay this interference pattern to be viewed through the eyepiece and then recorded as an interferogram. Later in the program a Vidicon TV camera replaced the "Extensometer" as a means of transferring the interference pattern.

Some time toward the end of the 60's, mirror-polishing operations were moved to a new facility at Building 601 on Lee Road. Final assembly testing moved to a large clean-room facility at Lincoln Plant.

APPENDIX B

The Formulas for Gambit-3 Revisited

For a long time I had suspected the cause of the disappointing performance of the early Gambit-3 missions lay in the original R-361 formula, but had no evidence. My suspicion was that the lens could be diffraction-limited on-axis but fall off so sharply in the field that the assessed performance could easily be lower by a factor of two. I believed this to be the case rather than the availability of high-resolution film.

In December of 1963, when Kodak made the proposal for Gambit-3, the only lens formula available was Kingslake's 120-inch R-360. I infer that the proposal included the chart, labeled "Ground Resolution vs Film Resolution", (Figure 9) which shows the projected ground resolution for a 154-inch focal length lens. There were no configuration layouts that I know of.

Since the R-361 formula was not issued until January 27, 1964, a complete layout could not have been available until February or March.

The R-361 design configuration was arrived at to avoid the unthinkable: which was to require a rectangular hole in the stereo mirror. When the Valley program put together the proposal for Gambit-3, Kodak was already having problems producing the much smaller Gambit stereo mirror. I participated in a later presentation, together with Cal Rogerson representing Quality Control, of the proposed optics configuration- the half-scale layout that stretched 7 feet across the front of the room. That lay-out showed the R-361 lens design with the aspheric 43 inch primary mirror, four element field flattener and the 58 by 46 inch elliptical stereo mirror- no hole through the center portion of the mirror.

Recently I was determined to follow up on the still nagging question: How well was the R-361 formula corrected off-axis? It occurred to me that Tropel might still have the design study report. A call put me through to Mark Westcott,[103] who informed me that the company had just cleaned house and all of their inactive files had been disposed of. After my briefing on the history of Tropel's important contributions to the nation's reconnaissance programs, he was kind enough to offer to run an analysis of the two formulas (on his own time). Not only was I pleased to once again establish the Tropel connection but also to see that, under the Corning umbrella, the science of lens design is still alive and well.

For me the comparison of these two Gambit-3 formulas, perhaps incomprehensible to anyone outside the community of optikers, confirms what I had suspected, that resolution of the lenses made to the R-361 formula would fall off sharply from the center of the image to the edge. However, the lenses should have performed quite well on-axis and over the central area.

In hindsight, one might question the decision (not to perforate the stereo mirror) that precluded the possibility of incorporating a Ross Corrector type lens formula. First of all, the Valley study program had not resolved the issue much before1964, the start of the program.

[103]Westcott, Mark R., B.S., 1970, Institute of Optics

The R-361 formula, issued January 27, 1964, was almost immediately incorporated into the optics configuration layout. The incredible schedule of events and the state-of-the-art in lens design certainly were contributing factors. Finally it was not until late in the 60's that the evolution of automatic lens design programs[104] could have produced the superb R-5 formula.

And so its only fitting that Tropel should be given the opportunity to evaluate R-361 formula, providing the missing link in the evolution of lens designs for optics in space.

KH-8 Gambit Optical Designs

Mark Westcott

Prescriptions adjusted before Design comparison:

1. Scaled R-361 prescription from EFL= 100 units, so that the Primary mirror diameter is 43.5 inch, same as R-5.
    - R-5 prescription was already presented at real scale.
2. Set both prescriptions to nearest modern Schott glasses.
    - Without original glass data, it is likely that both designs are degraded from their native design performance.
3. Set the object distance for both designs to 80 nautical miles.
4. Performed minor space adjust optimization to both designs.

Differences

Review of the ray intercept plots indicates that R-361 has more residual axial and lateral color and a little more coma than R-5.

R-5 has more high order spherical aberration and some lower ray issues, vs R-361.
Overall, R-5 has superior OPD.

Conclusions

Comparing MTF plots, R-5 outperforms R-361 at the ends of the image slit.

[104] The lens designer David Grey (credited with the original four-element Ross corrector formula) used what he called a differential correction lens design program "the Lens II program). John Buzawa at Tropel further refined the formula by adding a fifth element near the image plane.

Prescriptions adjusted before Design Comparison

1. Scaled R-361 prescription from EFL= 100 units, so that the Primary mirror diameter is 43.5", same as R-5
   - R-5 prescription was already presented at real scale
2. Set both prescriptions to nearest modern Schott glasses, as available on CODEV catalog
   - Due to modern "green" glasses, there are probably some differences in index, dispersion and partial dispersion
     - Without original glass data, it is likely that both designs are degraded from their native design performance
   - Used absolute glass indices, and vacuum in place of air
   - Used T= 20°C, although there was a tremendous temperature range in space. Athermalization is assumed
3. Set the object distance for both designs to 80 nautical miles
   - Not quite infinity: Mag= 0.00003
4. Performed minor space adjust optimization to both designs

Tropel Designs: First Order Summary

Dimensions in Inches

Designs were originally normalized to EFL = 100 units

Scale factor was classified & applied as needed

	R-361	R-5
Design Form	Aspheric Primary, 4-el Field Group	Aspheric Primary, 5-el "Ross" Field Group
Primary	General Asphere	Mostly Hyperbolic
EFL	160.1	173.7
EPD	43.5	
Spectrum	480 - 690 nm	
f/no	3.68	3.99
Altitude (nominal)	80 nautical miles	
Magnification	0.00003	
Angular FOV	$\pm$ 1.525°	$\pm$ 1.438°
Field slit length	8.53	8.79
Design date	Jan-64	Sep-68

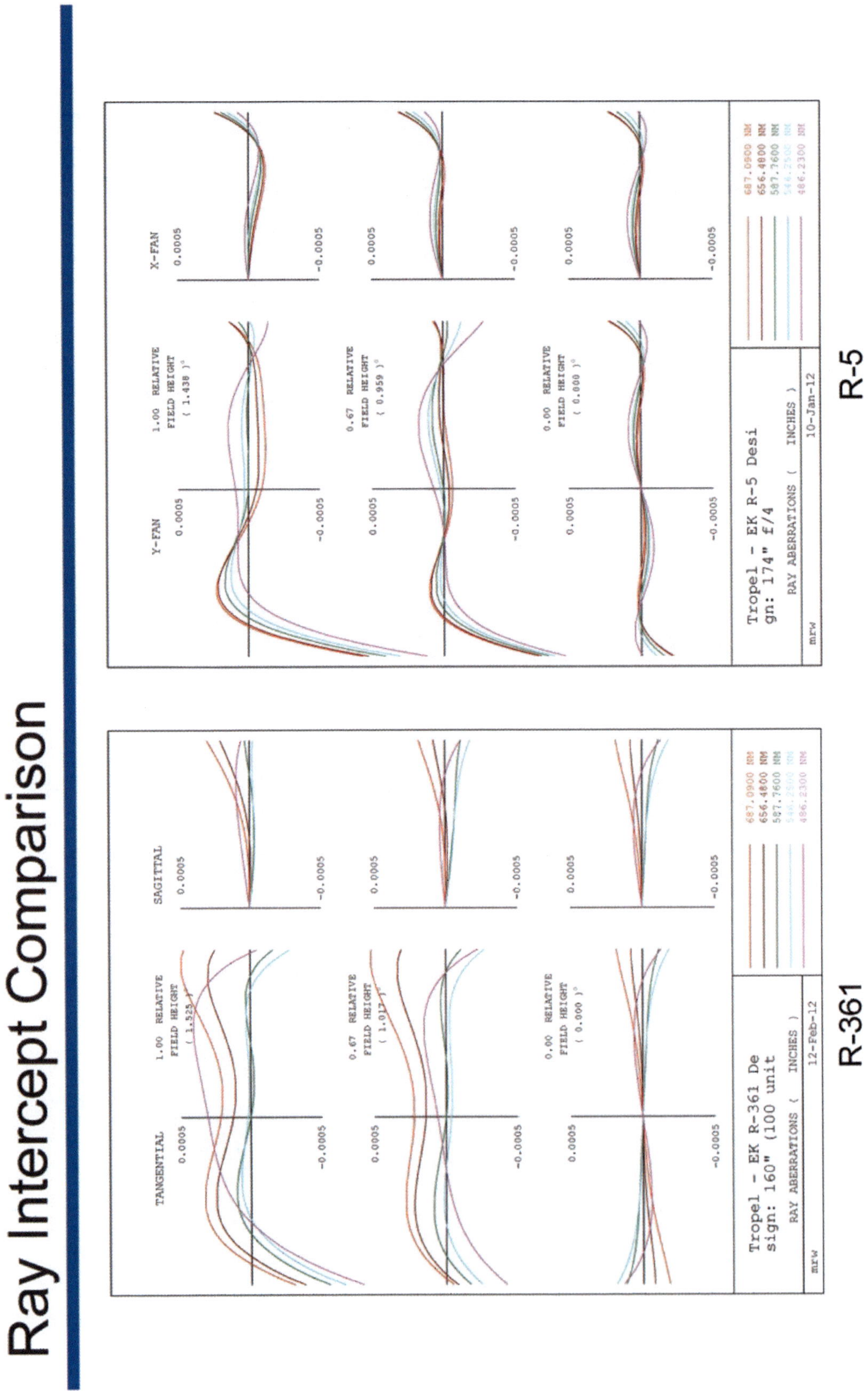
Ray Intercept Comparison
TANGENTIAL
SAGITTAL
1.00 RELATIVE FIELD HEIGHT (1.525)°
0.67 RELATIVE FIELD HEIGHT (1.017)°
0.00 RELATIVE FIELD HEIGHT (0.000)°
0.0005
-0.0005
Tropel - EK R-361 Design: 160" (100 unit
RAY ABERRATIONS (INCHES)
mrw
12-Feb-12
687.0900 NM
656.4800 NM
587.7600 NM
546.2500 NM
486.2300 NM
R-361
Y-FAN
X-FAN
1.00 RELATIVE FIELD HEIGHT (1.438)°
0.67 RELATIVE FIELD HEIGHT (0.959)°
0.00 RELATIVE FIELD HEIGHT (0.000)°
Tropel - EK R-5 Design: 174" f/4
RAY ABERRATIONS (INCHES)
10-Jan-12
R-5

OPD Comparison

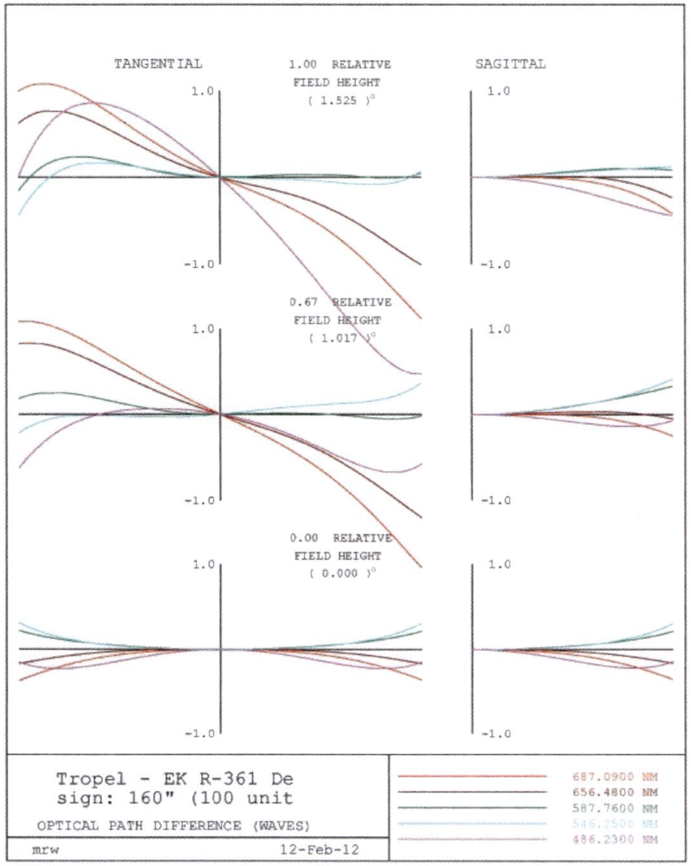

R-361

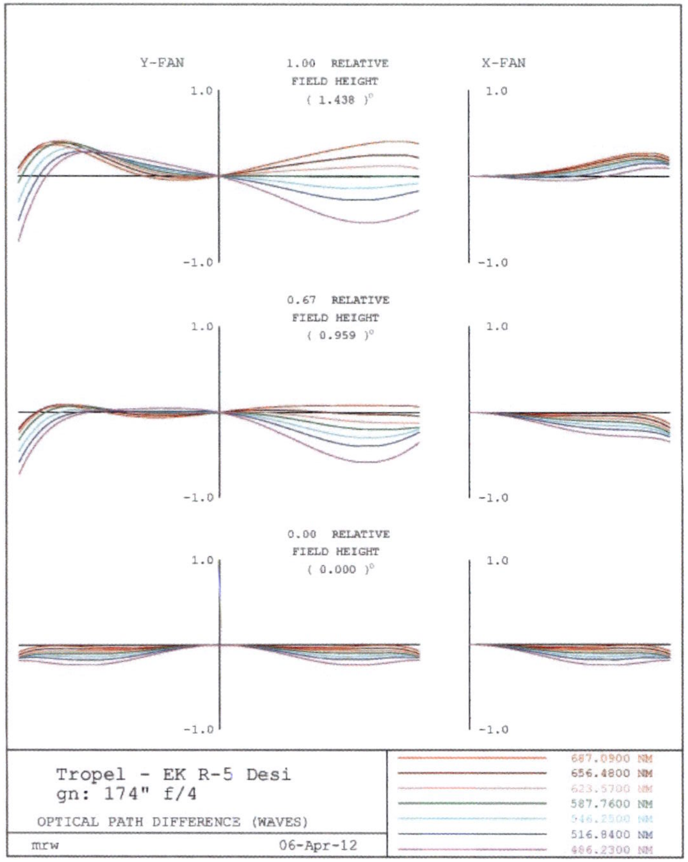

R-5

R-361 MTF

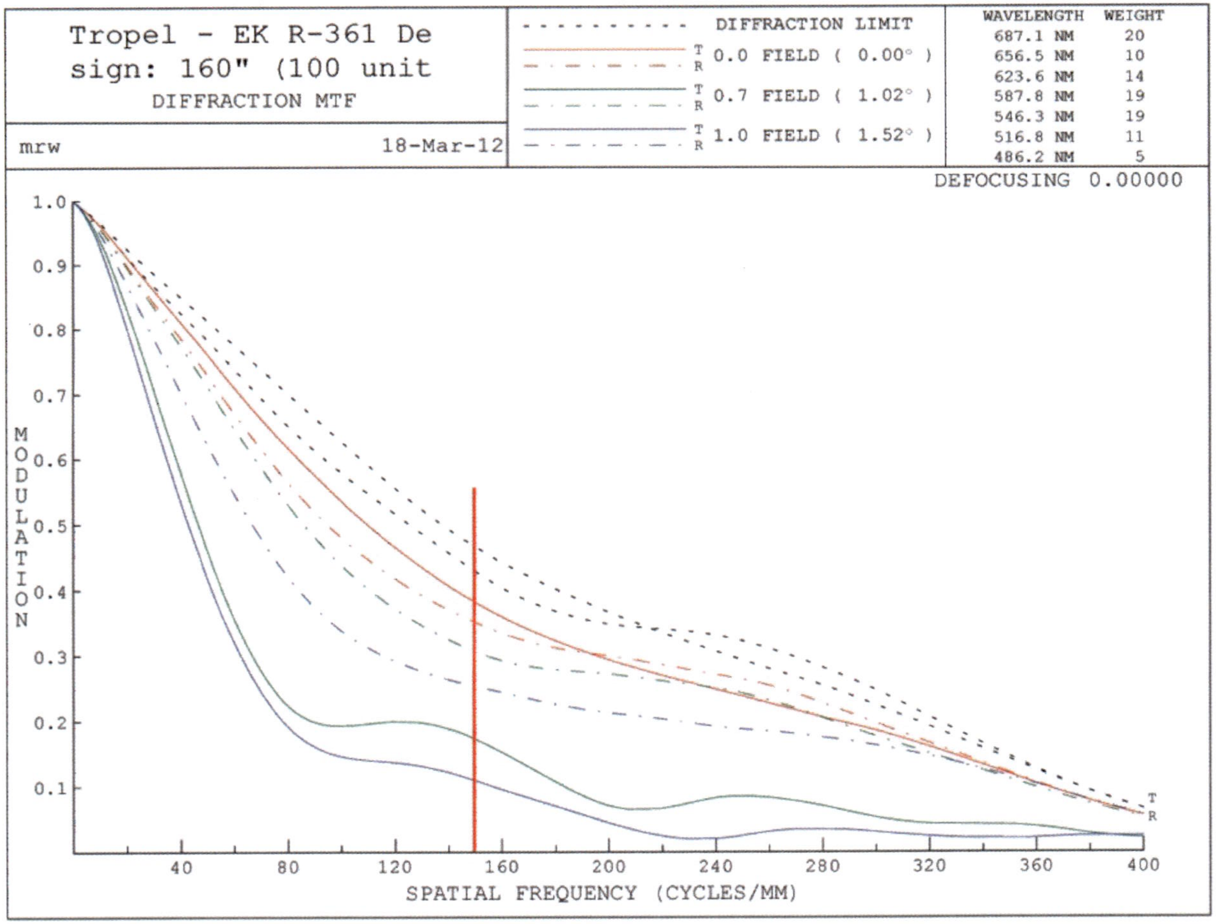
Tropel - EK R-361 De
sign: 160" (100 unit
DIFFRACTION MTF
mrw
18-Mar-12
DIFFRACTION LIMIT
T R 0.0 FIELD (0.00°)
T R 0.7 FIELD (1.02°)
T R 1.0 FIELD (1.52°)
WAVELENGTH WEIGHT
687.1 NM 20
656.5 NM 10
623.6 NM 14
587.8 NM 19
546.3 NM 19
516.8 NM 11
486.2 NM 5
DEFOCUSING 0.00000
MODULATION
1.0
0.9
0.8
0.7
0.6
0.5
0.4
0.3
0.2
0.1
40
80
120
160
200
240
280
320
360
400
SPATIAL FREQUENCY (CYCLES/MM)
T
R

R-5 MTF

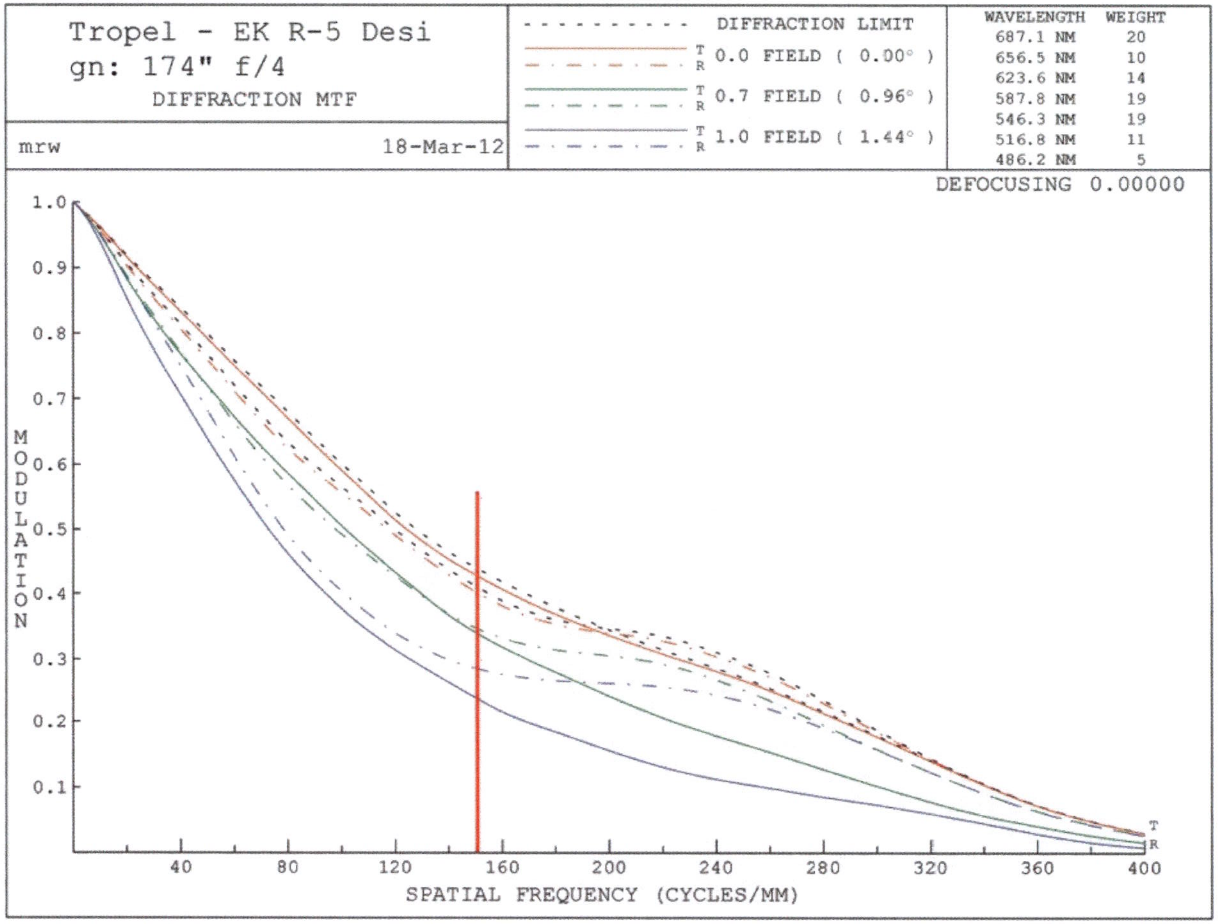
Tropel - EK R-5 Desi
gn: 174" f/4
DIFFRACTION MTF
mrw
18-Mar-12
DIFFRACTION LIMIT
T R 0.0 FIELD (0.00°)
T R 0.7 FIELD (0.96°)
T R 1.0 FIELD (1.44°)
WAVELENGTH WEIGHT
687.1 NM 20
656.5 NM 10
623.6 NM 14
587.8 NM 19
546.3 NM 19
516.8 NM 11
486.2 NM 5
DEFOCUSING 0.00000
MODULATION
1.0
0.9
0.8
0.7
0.6
0.5
0.4
0.3
0.2
0.1
40
80
120
160
200
240
280
320
360
400
SPATIAL FREQUENCY (CYCLES/MM)

IN RETROSPECT

I wish to acknowledge my indebtedness for the help offered during the final stages of publishing these memoirs. I should also acknowledge that I was inspired to incorporate the 26 years of my career working on secret government programs after reading Jeoffrey Richelson's book "America's Secret Eyes in Space."

In recognition of those who provided the encouragement and inspiration to continue the effort to publish:

Arne Jensen who succeeded in establishing the pipeline for obtaining approval to publish.

Dave Schaeffer for his oversight related to my involvement in classified programs at Kodak. His continued interest in this project provided encouraging support.

Bill Sabia whose artistic sense and experience in publishing as a subcontractor to Kodak provided not only an eye-catching cover design but also valuable editorial suggestions.

Jim Day, my son, for his encouraging support throughout this project and for his editorial assistance, proofing, typography design, page layout, photoshop work on all the images and for preparing the document for printing.

Sally Day, my wife of 65 years, for her quiet encouragement while suffering the endless hours of my absence on the Mac.

BIOGRAPHY

Pierce B. Day was born to China missionaries on October 25th, 1924. He received his early education in boarding schools, first at the Kuling American School in the mountains of central China, and at the Shanghai American School. His family returned to the U.S. in 1937 and settled in Wooster, Ohio where Pierce was placed in another boarding school, Peddie, for the remainder of his high school education. He was awarded the Genesee prize Scholarship at the University of Rochester in the fall of 1942, completed one semester, and then enlisted in the U.S. Air Force. After three years in the Air Force, he returned to the University of Rochester, Institute of Optics, taking courses in Physics and Optics, and graduating with a B.S. degree in 1948. He was employed by NACA, the predecessor to NASA, and returned to the U of R to study for a masters in optics, which he received in 1950, the year he started in the lens design department at Kodak.

Mr. Day is the sole or contributing inventor on ten U.S. patents, with the corresponding foreign patents that also issued. They are related primarily to possible copier configurations, however, two earlier inventions derived from work on the wind tunnel at NACA, a differential interferometer, and experimental work at Kodak resulting in the invention titled "Toric Surface for the Astigmatic Correction of a Shallow Dome."

Taken at the National Reconnaissance Office 50th Anniversary Gala, September 17, 2011 at the National Air and Space Museum, Washington D.C.

Printed in Dunstable, United Kingdom

91218892R00089